# Food Security in the Developing World

*Companion* website:

**http://booksite.elsevier.com/9780128015940**

*Food Security in the Developing World*
*John M. Ashley*

Resources for Professors:

**Global and Country Food Security Case Studies**

**Africa**
Case Study 1. Ethiopia: The Productive Safety Net Program (PSNP) (with particular reference to the Public Works component)
Case Study 2. Southern Africa and beyond: Conservation Agriculture

**Latin America & Caribbean (LAC)**
Case Study 3. Belize: Causes of undernutrition and poverty, and plans to address them
Case Study 4. Ecuador: Food security challenges from natural and man-made disasters
Case Study 5. Amazonia and La Plata Basin: regional food and other insecurities

**Asia**
Case Study 6. Lao PDR: causes of food insecurity in the mountainous north
Case Study 7. The Kuchi pastoralists of Afghanistan

**Global**
Case Study 8. Millennium Development Goals (MDGs), in particular MDG 1c relating to hunger and nutrition
Case Study 9. The food price crises of 2007-09 and 2011
Case Study 10. Urban slums: epicentres of deprivation and food insecurity

**References for companion site**

# Food Security in the Developing World

John M. Ashley

Senior International Consultant,
Geopolicity Inc., Dubai, UAE

AMSTERDAM • BOSTON • HEIDELBERG • LONDON
NEW YORK • OXFORD • PARIS • SAN DIEGO
SAN FRANCISCO • SINGAPORE • SYDNEY • TOKYO
Academic Press is an imprint of Elsevier

Academic Press is an imprint of Elsevier
125, London Wall, London, EC2Y 5AS, UK
525 B Street, Suite 1800, San Diego, CA 92101-4495, USA
50 Hampshire St., 5th Floor, Cambridge, MA 02139, USA
The Boulevard, Langford Lane, Kidlington, Oxford OX5 1GB, UK



ISBN: 978-0-12-801594-0

**British Library Cataloguing-in-Publication Data**
A catalogue record for this book is available from the British Library.

**Library of Congress Cataloging-in-Publication Data**
A catalog record for this book is available from the Library of Congress.

For Information on all Academic Press publications
visit our website at www.elsevier.com

Typeset by MPS Limited, Chennai, India
www.adi-mps.com

Printed and bound in the United States

*Publisher:* Nikki Levy
*Acquisition Editor:* Nancy Maragioglio
*Editorial Project Manager:* Billie Jean Fernandez
*Production Project Manager:* Lisa Jones
*Cover Designer:* Mark Rogers

**Cover Credits:**
Shutterstock #246931576: Malagasy farmer puddling his rice paddy
Shutterstock #175718696: Aerial view of Favela da Rochinha, the biggest slum in Brazil, on the mountain in Rio de Janeiro, with the skyline of the smart capital behind
Shutterstock #230865490: Corn cob in a parched field
Shutterstock #270676751: Farmers planting paddy rice, Thailand
Shutterstock #19148428: Tibetan nomads in the Himalayas

# Preface

This book is designed to provide an entry point primer in the subject of Food Security, something not done since the term first came into the public domain in the early 1970s. Books addressing the topic which are currently available contain learned debate on policy and research findings, analysis or collections of specialist papers. Despite their value, it is unfair to expect a relative novice to grapple with such without first having read a basic text, which defines and describes the subject's boundaries, contours and parameters. Without that first text the student may find it hard to develop the interest and confidence to proceed. This book offers to fill the space, with the intention to inform and facilitate beneficial change in the lives of the food insecure, who will never read the book.

Without having a comprehensive overview of the subject, those involved in national planning, and policy and strategy formulation may be underinformed, as also those involved in aid policy, and those whose role is to identify and design food security programs. Indeed, this book may serve as a textbook/field guide for a wide range of readers—undergraduate/diploma/vocational/adult education students and teachers; agricultural, rural development and planning ministries and departments; extension workers and nongovernmental organizations (NGOs); UN and other international organizations; bilateral and multilateral donor partners and development banks.

"Food Security" is not well understood by the general public. Too often the term equates only to food "availability". Often missing or insufficient is the imperative of seamless and holistic integration of food *availability* with economic, social and physical *access* to food, and the *nutritional, utilization and food safety* components, as also the imperative that all these parameters of achieving food security need to be accounted for on a simultaneous, present continuous and *stable* basis.

Chapter 1 provides an overview of the boundaries of the subject, and the topics addressed are expanded in subsequent chapters. Chapters 2–5 cover the manifestation and measurement of food insecurity, and the linkage between food security and nutrition security; the causes of this insecurity; and the mitigation of food insecurity of currently affected individuals and communities, and its future prevention for the most vulnerable. Chapter 6 addresses nine cross-cutting issues, followed by a final chapter with conclusions.

A companion website with 10 food security "case studies" from around the world is provided on the publisher's website, these being too lengthy for inclusion in the book itself. Each of these case studies refers to one or more of the key issues governing food security and nutrition security mentioned in the book, providing a more detailed context. Yet the book and companion website combined can still give only an overview of the all-embracing subject of "Food Security", which covers a galaxy of specialist fields. More than 250 references are cited where even more detail may be found. The list of acronyms and glossary of terms related to this book are both located on the companion website.

The author has addressed the subject as seen from the perspective of the developing world, this being the most pressing need. Yet, many countries have received not one mention, being omitted through shortage of space. It is hoped, however, that most categories and conditions of developing countries have been represented, dryland or monsoonal, land-locked or island State. The equivalent book from the perspective of the developed world would have given more space than allocated here to aspects of food security which may not be the primary concern of the developing world.

*Definition of the "developing world":* The concept of "developing countries" is relative, and far from clear-cut. Several international bodies, such as the United Nations Development Program (UNDP), World Bank and the International Monetary Fund (IMF) use different classifications, and an IMF working paper by Nielsen (2011) examines them. They are found lacking in clarity with regard to their underlying rationale, though they agree that between 20% and 25% of all countries in the world are "developed". Nielsen argues that there is no one generally accepted criterion for the definition, either grounded in theory or based on an objective benchmark, and that a country classification system based on a transparent, data-driven methodology is preferable to one based on judgment or *ad hoc* rules.

There are large and easily discernible differences in the standard of living enjoyed by citizens of different countries, with regard to household income and life expectancy for instance. While economists would likely agree that Burkina Faso is a "developing" country and that Japan is "developed", there are many countries "in between" which would be harder to classify as one or the other. This could suggest that a developing/developed country dichotomy is too restrictive, and that a classification system with more than two categories could better capture the diversity in development characterization across countries, with agreed numerical thresholds.

Classical economists determined that economic development was the nub of the dichotomy, the main proxy measure of development, referring to sustained increases in per capita real income. Yet there are social indicators too which are important measures of development, such as access to health facilities and education, sufficiency of nutritious food and nutritious diets, expected life span at birth, the prevalence of democracy and upholding of human rights by the State.

The "definition" of developing countries which has been adopted in this book references UNDP, which from 1990 adopted the Human Development Index (HDI), and produces an annual audit of countries according to a multiple proxy of indicators, called the Human Development Report (HDR). To capture the multifaceted nature of development, the HDI is a composite index of three indices measuring countries' achievements regarding its citizens' life expectancy at birth, education (two indicators) and income. Other aspects of development—such as political freedom and personal security—are also recognized as important, but lack of data across countries prevented their inclusion in the HDI.

The Index is subject to interrogation and amendment year on year. Each year, "development" is looked at from the perspective of the *people* within countries, rather than countries themselves, as the title *Human* Development Report affirms. Statistical Table 1, on page 160 of the 2014 HDR, shows a list of 187 countries, categorized into four groups corresponding to very high (1–49), high (50–102), medium (103–144) and low (145–187) human development. Countries in the last two categories are those for which the term "developing countries" and "developing world" most clearly resonates, and are the subject of this book. Sub-Saharan African countries constitute almost the whole of the "low" human development group.

The current author is aware that using national HDI as the "developing world" reference point excludes for consideration in this book certain subnational groups, such as those among the Inuit in northern Canada who still follow the nomadic fishing, hunting and/or pastoral livelihood. Canadian government data in 2008 showed that life span expectation of these people is 12–15 years shorter than people living in the rest of Canada. However, this is probably due to a dearth of medical and other social services in the remote and inhospitable north, and unhealthy overcrowded dwelling units, rather than to food or nutritional insecurity, or income poverty.

## REFERENCE

Nielsen, L., 2011. Classifications of countries based on their level of development: how it is done and how it could be done. IMF Working Paper. Strategy, Policy and Review Department, 49 pp.

# Acknowledgments

The author is grateful to his colleagues around the world who have provided assistance during the production of this book and companion website. He has given credits as necessary for generous permission from institutions and individuals to reproduce quotes, figures, tables, maps and photographs. Every effort has been made to trace the copyright holders of publications cited but if any have been inadvertently overlooked the publishers will be pleased to make the necessary arrangement at the first opportunity. He also thanks colleagues who agreed to read through parts of the text and offer comments—James Breen, Tariq Butt, Gino Henry and Jessica Lumala. Any errors or omissions throughout the book or companion website are of course the responsibility of the author. Nancy Maragioglio, Carrie Bolger and Billie-Jean Fernandez of Elsevier Food Science have kept the author on the straight and narrow, offering encouragement in the lonely task confronting every author, assuring this one that he was never alone!

The author would like to offer thanks to all those during his professional life who have taught him much of what he knows, from his early postgraduate days in Uganda until his most recent field assignment in 2014–2015, amongst the farmers and shepherds of the Jordan Valley. Finally, thanks are due to Trinity College, University of Cambridge, and Gianluca Olivieri of The International Management Group (West Bank and Gaza office) for providing the author with access to their facilities during the preparation of this book.

# ACKNOWLEDGMENTS

# Chapter | One

# Introduction

## 1.1 GENERAL SCOPE OF THE SUBJECT

### 1.1.1 Introduction

In 2014 the Food and Agriculture Organization (FAO) estimated that more than 800 million people are undernourished in terms of dietary energy intake. Another manifestation of undernourishment, micronutrient deficiencies, affects about 2 billion people. Each year more than 3 million children die of undernutrition before their fifth birthday.

Since the World Food Conference in Rome in 1974, the concept of food security has "evolved, developed, multiplied, and diversified" (Maxwell, 1996). Some 200 definitions of the term have been proposed, indicating a breadth of perspectives. However, the definition that has invariably been accepted now is that agreed at the World Food Summit (WFS) in November 1996, and this meaning is that endorsed in the current book, namely "*Food security* exists when all people, at all times, have physical, social and economic access to sufficient, safe and nutritious food which meets their dietary needs and food preferences for an active and healthy life".

Conversely, the FAO definition of *food insecurity* is "A situation that exists where people lack secure access for sufficient amounts of safe and nutritious food for normal growth and development and an active and healthy life". Food security, and its converse food insecurity, has rightly become a multidimensional index (Smith et al., 1993). The term "food insecurity" is not merely about the *concepts* of hunger and undernutrition, but the *experience* of these by those affected by or vulnerable to it, and the anguish of being unsure when one can next provide food for one's family. "Food insecurity" also inflames the passion of humanitarian and development partners who engage with it around the clock.

In the 1970s, food insecurity was understood (by those who did not have to bear its burden on their backs) as simply a measure of *availability* of food, evidenced in the various food crises and famines of the times. The Green Revolution, which significantly impacted developing countries in South America and Asia by the 1960s, certainly increased the food supply, yet hunger continued in much of the world. The realization dawned that many of those affected were unable to *access* the supply of food. Amartya Sen



Food Security in the Developing World. DOI: http://dx.doi.org/10.1016/B978-0-12-801594-0.00001-4

epitomized this in the statement "Starvation is the characteristic of some people not having enough food to eat. It is not the characteristic of there being not enough food to eat" (Sen, 1981).

### 1.1.2 The Multiple Dimensions of Food Security

The prevalence of *undernourishment* is a measure of dietary energy deprivation (hunger), though as a stand-alone indicator cannot capture the complexity and many dimensions of *food security*. The joint publication *The State of Food Insecurity in the World* in 2013[1] presented and analyzed a suite of indicators that address the four dimensions of food security: food availability, economic and physical access to food, food consumption/utilization and stability over time.

*Food security* is a term referring to the ability of a community, family or individual to be able to eat sufficiently, in terms of both quantity and quality, as prescribed by international standards of calorie, protein and vitamin intake. As indicated above, the term comprises several interrelated components (*food availability, access and also quality/safety/consumption-utilization*), all of which must be *simultaneously* achieved, to avoid the onset of "food insecurity". More recently the important component of *stability* of food supply over time has also been stressed, related to managing the risks to all of the above components due to abrupt shocks of an economic, conflict or climatic nature.

Food needs to be *available* in order to be accessed, from the family farm, the granary, the kitchen, and the local store or market. As a whole, over the 20 years up to 2014, food supplies have grown faster than population in the developing world, resulting in rising food availability per person (FAO Statistics Division, 2014). This is in spite of global food supplies exhibiting larger-than-normal variability in recent years, reflecting the increased frequency of extreme events such as droughts and floods (see Section 5.5.1).

Yet, availability alone is not enough to confer food security, for sufficient *access* to available food is often denied. This may come about as a result of *economic* constraints (poverty), and therefore the inability of an individual to purchase the food needed (if not receiving humanitarian aid). Improvements in economic access to food are reflected by reduction in poverty rates, which fell from 47% to 24% between 1990 and 2008 in the developing world as a whole. However, economic access to food, based on food prices and people's purchasing power, has fluctuated more recently.

Access to food may also be denied due to *physical* constraints, such as a road network which floods or gets washed away in the monsoon, cutting access to markets and raising the prices of food which can get through, or simply by living in a remote location. Also, the "separation wall" between Israel and Palestine, as a result of which people may be unable to access their resources (arable land to plant or harvest their crops, or the water for

[1] By FAO, the International Fund for Agricultural Development (IFAD) and the World Food Program (WFP).

irrigation or livestock), so that both food self-sufficiency and agricultural income generation are hopelessly compromised. The separation wall, by restricting personal movement, keeps many would-be workers away from employment opportunities, so they cannot earn enough to enable them to buy the food they need for themselves and the family.

In many parts of the world, there is a *social* or *political* barrier, such as that which prevents access to jobs (through class or partisan conflict, lack of opportunity for a good education, cultural norms), this keeping a person and her/his family poor and hungry.

In addition to availability and access, a third component of food security is consumption/utilization. Food may be both available and accessible in a community, but if an individual does not properly use or consume it, then that person is food-insecure, unable to benefit properly from either its availability or having access to it. Reasons underlying this could be compliance with societal social/religious norms, ill health, inability to cook it or store it safely (so it does not rot or become attacked by pests), the food being nutritionally inadequate, or the person unable to absorb its nourishment sufficiently (due to diarrhea or intestinal worms caused by poor hygiene or contaminated water).

Outcome indicators of *food utilization* convey the impact of inadequate food intake and poor health. *Wasting*, for instance, is the result of short-term inadequacy of food intake, an illness or an infection, whereas *stunting* is often caused by prolonged inadequacy of food intake, repeated episodes of infections and/or repeated episodes of acute undernutrition. Prevalence rates for stunting and *underweight* (a weight 15–20% below the "norm" for the age and height group) in children under 5 years of age have declined in all regions of the developing world since 1990, indicating improved nutrition resulting from enhanced access to and availability of food, although progress has varied across regions.

The fourth component of food security is *stability* over time of the food source, which relates to availability, access and predictability. Food price spikes can result from local, regional or international events, over which a given community has little or no control (see Case Study 9 on the book's companion website, http://booksite.elsevier.com/9780128015940) and clearly influence the stability of affordable food supply. Another cause of supply instability is cyclic seasonal events (such as alternating seasons of bumper and poor olive harvests in the West Bank of Palestine).

Where and when food *in*security is in evidence, at the levels of region, nation, community, households and/or individuals, not only is there an immediate human and social cost of incipient hunger and undernutrition but also detrimental long-term economics comes into play. A population can be seriously damaged through being ill-fed, in terms of irreversible loss of cognitive function in children, a discouraged and unemployed youth, vulnerability to disease and the cost of this to the health service and productive sectors. Causality is bidirectional—the food-insecure get sick, and the sick get food-insecure as they have not the energy to work, or look for work, to relieve that food insecurity. *A nation's human health is a strong predictor of the health of its economy*. It follows that there is a *huge macroeconomic cost to taking no action to relieve food insecurity in a sustainable manner—"we", the world community, cannot afford to do nothing about it.*

It may be seen from the above discussion that the term "nutrition" frequently occurs in the narrative. The concept of *nutrition security* is closely related to *food security*, the former being partly dependent on the latter. Without food security, it is impossible to have sustained nutrition security (see Section 2.1.3). Undernutrition of children under 5 years of age is especially concerning, particularly of children under 2 years of age, who are so completely dependent for their well-being on the mother (or caregiver in the absence of the mother). The United Nations Children's Fund stresses that the most irreversible damage caused by undernutrition globally happens during gestation and in the first 24 months of life (the first 1000 days), and this is the focus window addressed by many development partners (UNICEF, 2009). Beyond that, the period from 2 until 5 years is also regarded as a crucial window. As expressed to the current author in northern Nigeria "Once a child is five, he can beg, steal, scavenge and use a catapult" (to achieve a degree of food and nutrition security).

One in every five children in the developing world is undernourished, and poor nutrition is associated with half of all child deaths worldwide. Undernutrition in early childhood can lead to cognitive and physical deficits, and may cause similar deficits in future generations as undernourished mothers give birth to low birthweight babies. Undernutrition, through weakening the body, also increases susceptibility to and incidence of infections, and diminished response to vaccines. Undernutrition and infectious diseases are bi-causal and synergistic—undernutrition reduces immunological resistance and diseases deplete the body of essential nutrients.

One aspect of food insecurity which is taking hold in both developed and developing countries, for individuals and communities where food and wealth are readily available, is that of obesity. Such individuals are consuming more than "sufficiently" (as per the definition above) and are "malnourished" in the sense that their health is at risk from conditions such as heart and vascular disease, and diabetes. The frequent inactive and sedentary life style of such people means that excess calories are not burnt off, through laboring or exercise.

### 1.1.3 The Causes of Undernutrition

Nutritional problems in developing countries are normally caused by deficiencies or imbalances in dietary intake. These are conditioned by poverty, near-exclusive reliance on plant sources of nutrients, and high rates of infections. Undernutrition and infectious diseases further exacerbate poverty through lost wages, increased healthcare costs and impaired intellectual development that can reduce earning potential (Caulfield et al., 2006). Population subgroups at particular risk are the children and women of poor families. The root of undernutrition in early childhood is complex with a variety of direct and underlying contributors related to lack of balanced food, including insufficient breastfeeding and inadequate complementary foods; nutrient loss from gastrointestinal infections; chronic immune stimulation due to persistent parasitic intestinal infections; and, inadequate hygiene and sanitation, and contaminated drinking water (Psaki et al., 2012) (see Chapter 3: Causes of Food Insecurity).

### 1.1.4 Measurement of Food Insecurity

Of the three hierarchical components (availability, access, and utilization), availability is often measured through proxies at the population level, such as national agricultural output, while access and utilization are more often measured at the household and individual levels, respectively. While direct measures of food utilization exist, such as food frequency questionnaires, household food access has often been measured indirectly, through child anthropometry or agricultural productivity. Measurement of all three aspects of food insecurity has posed persistent challenges, such as the difficulty of measuring the impact of short-term shocks on household food access. Recent research, however, shows promise in the area of food access measurement, with the construction of simple household survey measures such as the Household Food Insecurity Access Scale. Low-cost and valid measures of household food insecurity are necessary to accurately predict the prevalence of food insecurity in response to changing conditions. Such measurements can then inform targeted interventions to diminish childhood morbidity and mortality (Psaki et al., 2012) (see Section 4.7).

Before proceeding with more detail in this book, it is important to realize that the term "food security" is not well articulated by the general public (see Box 1.1).

## 1.2 WORLDWIDE DISTRIBUTION OF FOOD INSECURITY

Food insecurity is clearly not evenly distributed across the world, currently or historically. There are pockets of food insecurity, examination of the distribution of which inevitably leads to a consideration of its causes (see Chapter 3: Causes of Food Insecurity) and progress in achieving Millennium Development Goal (MDG) 1, target 1c (see Case Study 8 on the book's companion website).

Global trends in hunger reduction mask disparities within and among regions. Latin America and the Caribbean (LAC) is the region that has shown the greatest progress in hunger reduction, with the prevalence of hunger reduced by almost two-thirds since the early 1990s. As a whole, by 2014 LAC had already reached the MDG hunger target and was very close to meeting the more stringent WFS target. Government-led efforts combining support for production with social protection have been supported by much wider commitment—societies have

**Box 1.1 Widespread Unawareness of the Term "Food Security"**

Having briefly described above the scope of the term "food security", it is nevertheless a term not encompassed in the lexicon of the general public. The current author has frequently been asked what it means, and if it refers to terrorists or cash transit vans? Some 200 definitions of the term occur in the literature, so perhaps the confusion is not surprising. Rather more surprising is the incomplete grasp of the term by some development agency officers.

In 2005, when involved in formulating Palestine's Food Security Strategy, the author visited the Representative Office of a major donor in Ramallah to enquire what food security initiatives were being funded from that office. He was assured that there were no such initiatives whatsoever. When pressed to say what development initiatives *were* being followed, the development officer started by saying that there was a new port facility being planned for Gaza. "Thank you very much. That's Food Security. And what else is your office funding please . . .?". There then followed several other examples of initiatives which would lead to improved food security for Palestinians.

Another example from the same country came to light in 2014, when a different donor office was asked if it were supporting any agribusiness initiatives (which directly relate to food availability, and accessibility via increased profitability of agriculture). A total absence of any such initiative was acknowledged. When confronted with documentary and pictorial proof that this was not the case, the officer concerned was embarrassed. It transpired that help given to women's associations to increase their business acumen and facilitate microcredit was classified in embassy parlance as "social cohesion" initiatives, though the women were exclusively engaged with value-adding processing of primary agricultural products.

Indeed, it is more challenging to think of development initiatives which would *not* lead to increased food security for a given country or community. At the personal level too, so many things each of us does each day are with the intention of improving our personal and family food security, without that term ever being articulated.

decided to end hunger and parliaments are taking responsibility for that. National efforts have been propelled by the strong commitment of the region as a whole, which became the first region to commit to the goal of zero hunger (see Section 2.1.1) by adopting in 2005 the Hunger-Free Latin America and the Caribbean Initiative 2025, the initiative aimed at reducing the impact of chronic undernutrition in children to below 2.5% in all the countries of the region by 2025 (see Section 6.6.3). This commitment is being strongly monitored, and has been reaffirmed by the region's leaders at recent Summits of the Community of Latin America and the Caribbean States (CELAC).

The 2014 report *The State of Food Insecurity in the World* observes that the size of Asia makes it a region of extremes (FAO, IFAD and WFP, 2014). Though 217 million Asians have overcome hunger since 1990–92, two-thirds of the world's hungry live in Asia. While the MDG hunger target was already achieved in Eastern and South-Eastern Asia by 2014, hunger prevalence in Southern Asia had declined *in*sufficiently to meet MDG 1c.

While Northern Africa has had a consistently low prevalence of hunger at less than 5%, in sub-Saharan Africa one in four people remain chronically hungry. Reversing this trend is a huge challenge, to embody the growing political will in the region shown by the commitment made at the June 2014 African Union Summit to end hunger by 2025.

At a lower level than regional and national disparities are differences relating to degrees of poverty, and predisposition of communities to disease. These are both urban and rural challenges. For instance, in the marginalized fringes of wealthy town centers in the developing world, there are usually slum settlements. The number of slum dwellers is increasing, both in absolute terms and as a proportion of the world's population. For instance, in Nairobi, Kenya, there are 2.5 million people living in some 200 slum settlements. These represent 60% of the total population of Nairobi, though occupying just 6% of the land in the city. Kibera slum alone houses almost 1 million of these slum dwellers in just one square mile, it being the largest slum in Africa and one of the largest in the world. Food and nutrition insecurity therein is rampant (see Case Study 10 on the book's companion website).

## 1.3 NUMBER OF FOOD-INSECURE PEOPLE

Estimates from *The State of Food Insecurity in the World* report of 2014 show that, since 1990–92, the prevalence of undernourishment has fallen from 18.7% to 11.3% in 2012–14 for the world as a whole, and from 23.4% to 13.5% for developing regions. Since 1990–92 the number of hungry people has fallen by over 200 million. Despite this progress, however, the number of hungry people in the world is still unacceptably high: at least 805 million people, 30 million of whom are in the African Sahel region and the Horn of Africa (this figure was revised downward to 795 million in the 2015 edition of the FAO/IFAD/WFP report).

Stated somewhat differently, the figures are even more arresting—"one in nine people in the world will go to bed hungry tonight". And many of the 795 million do not have beds—they sleep where they can, in shop doorways or sewers, in cities like Lima, Vientiane, Kathmandu, Sana'a, Mogadishu and Lusaka. These children sleep not only hungry, but dirty, abused and unloved, with no expectation that "tomorrow" will be any different. For one hungry boy at least though, "tomorrow" *was* different. On June 6, 2011, when former south Korean street child Choi Sung-bong had a chance to sing on the show "Korea's Got Talent", the world's media had a sensational story to tell. A *youtube* clip of him had received 142,841,189 views as of May 30, 2015.[2]

As the EC (2009) cogently points out, "despite the image conveyed by the media, fewer than 8% of the people affected by hunger are victims of food emergencies. The world's 963 million inhabitants[3] who are hungry rarely make the headlines. Nevertheless, there are more of them today than there are people living in the United States, Japan and the European Union combined".

---

[2] www.youtube.com/watch?v=tZ46Ot4_ILo.

[3] In 2015 estimated at 795 million.

The day after World Food Day in 2013, the *Al Jazeera* website ran an item it headlined "Why are so many people still hungry?".[4] It stated that, according to a report from the United Nations, there has been a decrease in the number of starving people around the world. Between 2011 and 2013, a total of 827 million people were hungry in developing nations, this being 169 million fewer than the same statistic 20 years previously. More than 60 countries had reached, or were expected to reach, their Millennium Development hunger targets, and the Global Hunger Index released in mid-October 2013 showed that overall, sub-Saharan Africa has a better score than South Asia. That is a twist on the thinking that Africa has the worst levels of hunger and it is a show of progress for the continent.

The *Al Jazeera* article quotes Alexandre Meybeck of FAO "This is very good news, and especially a reason to do more. One should not forget that there are more than two billion people who are malnourished, lacking essential vitamins, and minerals which are absolutely essential for good health, and it is especially important for children. So it is not only the global figures of hunger to keep in mind, but also what is called hidden hunger—people who are not fed as they need to be". The article suggested that there was little to celebrate as some parts of Africa were then experiencing the worst drought in decades. About 1.5 million people in southern Angola, and more than 778,000 in northern Namibia, did not have enough food or water. If nothing were done, a severe famine would be on its way, according to UN reports. For many in Africa, and in other parts of the world, undernourishment is already a reality. Zambia, Mozambique and Ethiopia are of the biggest concern. In Asia, Laos and North Korea stand as some of the most undernourished countries, and even India, with its population of more than 1.2 billion, has a major issue with food security. The world's population is expected to reach more than 9 billion by 2050 and much of the planet's arable land is already in use. Population growth of course will continue beyond 2050. The outcome will be rapid growth in need for food, in terms of both quantity and quality.

There is a shifting demographic issue too. Exponential population growth is being accompanied by unprecedented rates of urbanization, particularly in developing countries. Fifty percent of the world's population live in urban areas, which means that they no longer depend on producing at least some of their own food but rather on buying all of it, so its affordability becomes more of an issue for them.

## 1.4 SHORT-TERM FOOD INSECURITY COMPARED WITH LONG-TERM STRUCTURAL FOOD INSECURITY

Food insecurity may be of a relatively *short-term transitory nature*, due perhaps to a missed rainy season for a year or two. If that part of the country or

---

[4] Retrieved from http://www.aljazeera.com/programmes/insidestory/2013/10/20131017102112795784.html.

region affected is not adequately prepared for the challenge, there can be serious consequences for those who live there. Coping strategies become exhausted, and *structural* poverty ensues (see Section 4.1). The northern Ethiopian famine in 1983–85 is one such case, in which the government of the day, for political reasons, did not communicate the crisis to the outside world until it had already claimed thousands of casualties, and investigative television journalism had blown its cover. The famine's primary cause was government policies to counter "insurgencies", and latterly drought.

By contrast, food insecurity may be more *long-term*. Such can arise from deleterious climate change, a food price crisis or conflict. During the two recent civil wars in Liberia, West Africa, for instance (1989–96 and 1999–2003), as well as those fleeing to neighboring countries as external refugees, there was massive internal (within-country) displacement of people, so that even had they planted food crops they would not have been there to harvest them or their perennial cash crops (rubber, cocoa and coffee). While they were away from their homes, their livestock were stolen or eaten by one of the militia rampaging the country, and the rubber trees tapped for the latex by gangs in an unprofessional way, such that the productive potential of trees was ruined and vast areas of rubber plantation were laid to waste as an economic venture. With the productive base decimated, and a generation of children not having received an education, *structural poverty* ensued. Many of the original inhabitants who fled the country have still not returned to Liberia, being unable to trust their neighbors and afraid of what may await them in future. Livestock numbers remain minimal, as families still do not have the confidence to build up the flocks and herds.

The more recent escalation of internal conflict in Syria has been going on since 2011, in part a domino effect downstream of the "Arab Spring" in North Africa. Each day its complexity increases as does the death toll. Opposition to the Assad regime is splintered, with internecine skirmishes between rivals (with their international backers), and millions of Syrians becoming internally displaced or fleeing to neighboring countries or further afield, some drowning in the Mediterranean as they endeavor to reach Europe. As weapons inspectors strove to disable the Syrian government's arsenal of chemical weapons and their ability to use them, diplomats outside of the country struggled to achieve a negotiated peace against the background of conflicting narratives by powerful parties with incompatible vested interests. How is it possible to know, or even guess, how long this particular phase of personal insecurity and associated food insecurity will continue in the country, and therefore to know whether the situation can be classified as transitory or long-term? Only in retrospect will it be known. At the moment, the answer appears to be "long-term".

Support is needed for vulnerable populations because even a transitory crisis can trigger chronic food problems, as assets are quickly depleted and livelihoods undermined. Food insecurity is particularly hard to tackle in complex ongoing crises and in the fragile transition to stability. During a crisis, fragile states may lack the capacity or institutional framework to implement long-term food security solutions—a situation that may be compounded by poor governance, conflicts, man-made disasters, malaria, measles, HIV/AIDS, Ebola and other diseases.

## 1.5 FLASHPOINT GRAIN DEFICIT AREAS COMPARED WITH TRADITIONAL GRAIN SURPLUS AREAS

Regions of the developing world which are prone to drought and/or flooding, the economies of which are still modest, are the most vulnerable to food insecurity, the risk factors for food availability and economic access being significant. Bangladesh with its heavily populated low elevations and tumultuous Brahmaputra and other rivers in the monsoon is at high risk from flooding on a recurrent basis, increasingly so because of the amount of silt being deposited in the river bed, caused by forest felling on the slopes above, and subsequent soil erosion. As for drought, countries in the Sahelian region of Africa, such as Mali and Mauritania, are at high risk of food insecurity. As areas of NE Nigeria become drier, so farmers are replacing their sorghum staple by the more drought-tolerant millet. A relatively new powerful force on the scene is climate change, as discussed in Section 6.9, this set to create an extension of the drylands and inundation of coastal cities (HLPE, 2012). Both will clearly have a huge impact on food and nutrition security of the communities affected, and the global cereal and oilseed markets. Already, 2.5 billion people live under dryland conditions.

The relentless population increase in many parts of the world puts pressure on natural resources, land and water in particular, with the threat of loss of soil fertility and surface erosion. Mentioned elsewhere in the book (see Section 6.4) is the extra million people per year in Kenya, much of which is arid or semi-arid land, and already challenged to feed its citizens. There are almost 10 million "climate refugees" around the world who have been forced to leave their homes because of lack of rain and desertification, or rising sea levels. Moreover, countries that are absorbing refugees from violence or discrimination in neighboring States also have their national food security compromised. For instance, in 2015, in addition to many refugees fleeing from South Sudan and DR Congo, Uganda is hosting thousands more from political strife in Burundi. The United Nations High Commissioner for Refugees (UNHCR) and the Government of Uganda struggle to cope with providing the food needed for people who cannot produce their own.

Many oil- and gas-rich states of the Persian Gulf produce no more than a few percent of the food they need, yet are (currently) sufficiently wealthy to be able buy in what they need, even in periods of food price hikes.

In terms of tonnage, the main rice producers in the world in 2012 according to the FAO Statistics Office (FAOSTAT) are China, India and Indonesia; for wheat, the European Union, China, India, USA and the Russian Federation dominate; and, USA, China and Brazil dominate corn (maize) production. In terms of countries which sell onto the open market, from which food deficit countries could buy in times of crisis, the picture is rather different—according to various data sources, dominant exporters of rice in 2013 were India, Thailand, Vietnam, Pakistan and USA (only one-third the amount of India); for wheat, export was dominated by the European Union, USA, Canada, Australia and the Russian Federation; for corn (maize), the dominant exporters were USA, Brazil (only half as much as USA), Ukraine and Argentina (see Section 5.5.1 for global grain reserves).

## 1.6 FOOD AS A RIGHTS ISSUE

Freedom from Want, including need of food, was one of the Freedoms mentioned by President Roosevelt in his "Four Freedoms" speech in 1941. After Roosevelt's death and the end of World War II, his widow Eleanor often referred to the four freedoms when advocating for passage of the United Nations' Universal Declaration of Human Rights. Mrs Roosevelt participated in the drafting of that Declaration, which was adopted by the United Nations in 1948. In Article 25, it states, "Everyone has the right to a standard of living adequate for the health and well-being of himself and his family, including food". This provided a reference point for human rights legislation that followed, though the Article not itself a binding international legal instrument.

Freedom from hunger is such a basic human right, yet more than 10% of humanity remains hungry. There are two main global initiatives to reduce hunger:

- the WFS of 1996 in Rome, its commitments, and follow-up
- the MDGs of the UN Millennium Summit in 2000, and their follow-up

Both the WFS and the MDGs recognize the importance of food to alleviate hunger and its importance to human beings, though not addressing it in the context of a legal right. However, as major international efforts they can facilitate the human right to food. Right to Food advocates recognize that this right cannot be achieved overnight in developing countries, and the importance of the concepts of progressive implementation and voluntary guidelines.

There is more than enough food in the world to feed everyone, but the number of people affected by hunger and undernutrition is still unacceptably high with disproportionate impacts on women and girls. Reversing this state of affairs must be a top priority for governments and international institutions. A report in 2014 by the Institute for Development Studies urges that responses must treat food insecurity as an equality rights and social justice issue (IDS, 2014).

Most importantly, food and nutrition insecurity is a gender justice issue. Low status and lack of access to resources mean that women and girls are the most disadvantaged by the often-inequitable global economic processes that govern food systems, and by global trends such as climate change. Evidence shows strong correlations between gender inequality and food and nutrition insecurity—for example, despite rapid economic growth in India, thousands of women and girls still lack food and nutrition security as a direct result of their lower status compared with men and boys. Such inequalities are compounded by women and girls' frequent limited access to productive resources, education and decision-making processes, by the "normalized" burden of unpaid work—including care work—and by the endemic problems of gender-based violence (GBV), HIV and AIDS.

At the same time, women literally "feed the world". Despite often limited access to either local or global markets, they constitute the majority of food producers in the world and usually manage their families' nutritional needs. They achieve this despite entrenched gendered inequalities and increasing volatility of food prices. Yet their own food security and nutrition needs—and often those of their daughters—are neglected at household level, where discriminatory social and cultural norms prevail.

There needs to be a new, gender-aware understanding of food security, on the basis that partial, apolitical and gender-blind diagnoses of the problem of food and nutrition insecurity are leading to insufficient policy responses, and the failure to realise the right to food for all people. Showcasing effective and promising existing strategies, the IDS report of 2014 *ibid.* proposes that in order to truly achieve food security for all in gender-equitable ways, responses need to be rights-based, gender-just and environmentally sustainable (the gender perspective is discussed further in Section 6.2).

## 1.7 INVESTMENT AS THE APEX SOLUTION TO FOOD INSECURITY

When situations of large-scale human food insecurity arise around the world, the combined humanitarian aid efforts undertaken have undoubtedly saved many lives, yet they are not sustainable in the medium- or long-term. Also, there is no "development" component to traditional humanitarian aid, enabling more food to be produced. At best it is a stop-gap measure, while at worst it can create dependency and undermine the will of people to become independent of food aid. The ideal should surely be for humanitarian aid to be phased out, in favor of development initiatives that enable the people concerned to grow more food and/or create wealth enabling them to buy (access) the food which is available, and/or to provide for a permanent "safety net".

Such latter interventions may take the form of strategically placed grain stores (see Section 5.5.1), or transformation of the environment to make it more productive (see Case Study 1 on the book's companion website). Indeed, many international agencies are actively involved in strengthening a country's physical and social assets, which goes some way to providing a food-secure future. Such a transition from humanitarian aid to development is termed LRRD—Linking Relief, Rehabilitation and Development.

The 2014 *Human Development Report—Sustaining Progress: Reducing Vulnerabilities and Building Resilience*—looks at these two concepts which are both interconnected and vital in securing human development progress. Since UNDP's first global HDR in 1990, most countries have registered significant human development. The 2014 Report shows that overall global trends are positive and that progress is continuing. Yet, lives are being lost, and livelihoods and development undermined, by natural or human-induced disasters and crises.

FAO has estimated that agricultural production needs to increase by 60% between 2005 and 2050 to enable the estimated nearly 10 billion people then to be adequately fed. Already, the increasing population pressure on land and water is reducing agricultural productivity and environmental integrity. Climate change will likely intensify that pressure in the developing world, adding to the difficulty of reconciling food needs with production potential. *Food production and use systems are needed which are equitable and sustainable in all dimensions—economic, social and environmental.*

To meet this challenge, agricultural production in many developing countries may need to double, to counterbalance those countries in which a 60% increase is not possible. In 2009, the United States Agency for International Development (USAID) calculated the net investment required to support this expansion in agricultural output and associated agricultural productivity, at $83 billion annually on average. This figure includes investment needs in agricultural and necessary downstream services (such as processing and storage facilities), but does not include agricultural investment for roads, electrification and large-scale irrigation projects. Current investment flows are significantly insufficient to achieve such agricultural and infrastructural development goals.

As 70% of the world's poor live in rural areas and produce most of the world's food, smallholder agriculture needs to receive particular focus for such investment. This in turn needs a supporting enabling environment in terms of policies, and access to markets and finance. Sustainable food systems need to ensure not only food security for all and environmental protection, but the creation of decent employment opportunities whereby those who do not grow the food can afford to buy it, thereby also stimulating multiplier effects. For all this to happen, smallholders need access to functional local and regional markets, supply chains and means of value-adding processing. Food and nutritional security, capacity building, jobs, growth and finance are all inextricably linked, within the context of removing poverty and food insecurity while increasing global food supply (see Section 5.5.2 for further discussion on investment).

A report by ActionAid USA (2013) cites a growing body of experience at the local and regional levels which demonstrates the lasting value of investments in climate-resilient smallholder farming and sustainable agricultural methods (for instance, as in the "New Alliance for Food Security and Nutrition" initiated by the G8), rather than investments in large-scale, input-intensive agricultural development, driven by private sector-led projects.

## 1.8 UN ORGANIZATIONS WHICH DIRECTLY ADDRESS FOOD AND NUTRITION SECURITY

FAO: The Food and Agriculture Organization works to alleviate poverty and hunger by promoting agricultural development, improved nutrition and the pursuit of food security. The Organization offers direct development assistance; collects, analyses and disseminates information; provides policy and planning advice to governments; and, serves as an international forum for debate on food and agriculture issues.

IFAD: The International Fund for Agricultural Development is dedicated to eradicating rural poverty in developing countries. IFAD focuses on country-specific solutions, which can involve increasing poor rural people's access to financial services, markets, technology, land and other natural resources.

WHO: The objective of the World Health Organization is the attainment by all peoples of the highest possible level of health. Health, as defined in the WHO Constitution, is a state of complete physical, mental and social well-being, not merely the absence of disease or infirmity. WHO is the directing and coordinating authority on international health interventions.

UNICEF: The United Nations Children's Fund is the leading humanitarian development agency working globally for the rights of every child—these rights include safe shelter, nutrition, protection from disaster and conflict, prenatal care for healthy births, clean water and sanitation, healthcare and education.

WBG: The World Bank Group is the world's largest source of development assistance and uses its financial resources, highly trained staff and extensive knowledge base to help each developing country onto a path of stable, sustainable and equitable growth in the fight against poverty.

UNDP: The United Nations Development Program helps countries in their efforts to achieve sustainable human development by assisting them to build their capacity to design and carry out development programs in poverty eradication, employment creation and sustainable livelihoods, the empowerment of women and the protection and regeneration of the environment, giving first priority to poverty eradication.

UN Women: UN Women works toward the elimination of discrimination against women and girls; empowerment of women; achievement of equality between women and men as partners and beneficiaries of development, human rights, humanitarian action, and peace and security.

UNHCR: The United Nations High Commissioner for Refugees is mandated by the UN to lead and coordinate international action for the worldwide protection of refugees and the resolution of refugee problems. UNHCR's primary purpose is to safeguard the rights and well-being of refugees. UNHCR strives to ensure that everyone can exercise the right to seek asylum and find safe refuge in another State, and to return home voluntarily.

UNRWA: The United Nations Relief and Works Agency for Palestinian Refugees in the Near East provides direct relief and works programs to

Palestine refugees, in order to prevent conditions of starvation and distress and to advance conditions of peace and stability.

WFP: The World Food Program is the food aid arm of the UN. Food aid is one of many instruments that can help promote food security. The policies governing the use of WFP's food aid are oriented toward the objective of eradicating hunger and poverty. The ultimate objective of food aid should be the elimination of the need for it.

OCHA: The mandate of the Office for the Coordination of Humanitarian Affairs of the UN includes coordination of humanitarian response, policy development and humanitarian advocacy. OCHA carries out its coordination function primarily through the Inter-Agency Standing Committee (IASC) chaired by the Emergency Relief Coordinator. Participants include all humanitarian partners, from United Nations agencies, funds and programs, to the Red Cross Movement and NGOs. IASC ensures interagency decision-making in response to complex emergencies. These responses include needs assessments, consolidated appeals, field coordination arrangements, and the development of humanitarian policies.

## 1.9 HISTORY OF FEEDING THE WORLD FROM 1945

Conway (2012) set a challenge for us all in his compelling book entitled "One Billion Hungry: Can We Feed the World?". He believes that we have the collective ability to do so, with the proviso that we use technology paired with sustainable practices and strategic planning. This optimism is set against the historical record of addressing global hunger, undernutrition and food insecurity in their various manifestations during the 1980s and 1990s, being a record of both qualified success and unjustifiable failure (Kracht and Schulz, 1999). Influential players on the world stage, including the UN and other international organizations, have learned from their early efforts at trying to address food and nutrition insecurity in the developing world, though it has taken a long time to "get their act together", and there is some way to go.

The classic treatise on how the world has endeavored to assure its food security on a global scale since 1945 is by Shaw (2007), almost 500 pages of those successes and failures, and analysis. The book covers in detail aspects such as commodity agreements and global food reserves, early warning systems and role of UN agencies. It details the early pioneering work by FAO, the food crises of the 1970s and how the world responded up to the 1990s, and the many international follow-up conferences. The book also probes the interlinkage of food insecurity and poverty.

After the world food crisis of the early 1970s and the 1974 World Food Conference, global efforts to tackle food insecurity centered on increasing food production, stabilizing food supplies and using developed world "food surpluses", despite efforts of the World Food Council to broaden this focus. The International Labor Organization World Employment Conference of 1976, with its concept of "basic needs", and the economic concept of Sen's food "entitlement" (means to access commodities), led to an acknowledgment of the need to assure *access* to food by the hungry poor, thereby expanding the understanding of food security from agricultural supply into the broader arena of poverty and development. Addressing in a coherent manner the raft of interlocking determinants of food insecurity became a priority.

Part IV of Shaw's book provides a look into the future and is both daunting and inspiring. One aspect covered is oversight governance of what is to be done to address food and nutrition security at the global level, including how to improve institutional coherence. With so many multilateral, bilateral, NGOs and international organizations involved, food security has become "everybody's concern and so, in reality, no one's concern". Developing common and coherent policies, priorities and programs to attain world food security has been and continues to be a considerable challenge.

Several phases of food security policy and practice were detected by Shaw. The first phase was in the 1970s, following the world food crisis and the 1974 World Food Conference, when the objective was to establish a global food security system. This phase was followed by competing rhetoric during the first half of the 1980s, with Sen's food "entitlement" concept on the one hand and the IMF and World Bank structural adjustment programs on the other. Analysis of the complex causes of food insecurity was renewed in the second half of the 1980s, in the face of the Africa famine of 1984–85, and the publication of *Adjustment with a Human Face* by UNICEF in 1987, and *Hunger and Public Action* by Drèze and Sen in 1991. A series of international conferences in the 1990s emphasized poverty as the major cause of hunger and undernutrition.

This broadening of the concept of world food security, with its multidimensional aspects, in many ways mirrors the contemporary evolving views on development theory and practice. The 1960s saw the recognition of the importance of the human factor in development. Sustained and equitable development depended not on the creation of wealth itself but on the capacity of people to *create* wealth. This led in the 1970s to an interrogation of the paradigm of an automatic "trickle-down" of the benefits of development to the poorest. The 1980s witnessed world economic recession, the poor suffering the most as a result of economic adjustment programs imposed on developing countries by international financial institutions. The 1990s and beyond has underlined the centrality of the human factor, enshrined in the annual *Human Development Reports* of UNDP, and the mainstreaming of employment as a gateway to assured access to food, and mainstreaming gender and marginalized group priorities in development initiatives (Emmerij et al., 2001; Jolly et al., 2004).

Over recent decades, powerful commercial forces have entered the world food system, including the emergence of large multinational food corporations which have largely taken control of the food chain (Maxwell and Slater, 2003; McDonald, 2010). Urbanization in both developing and developed economies has led to changing patterns of food consumption, emergence of the fast food industry and supermarket chains, with related saturation advertising. Obesity, not undernutrition, has emerged as a major killer in the developed world and in parts of the developing world too, due to these changed food habits and more sedentary work practices and life style. Government import and pricing policies, changes in the relative prices of food commodities, increasing income, transport and logistical improvements and changes in fuel costs can influence both food habits and food security. International trade and finance tends to be dominated by a small number of developed nations (Uvin, 1994), and this has led to political unrest and development of the food sovereignty movement (see Section 6.8).

Food reserve management was an important component of food security policies in the 1960s and 1970s. However, in the following decades their role diminished, as they were often considered a costly and inefficient mechanism, especially during the structural adjustment reforms spearheaded by the Bretton Woods Institutions. Nonetheless, during the 2007/08 food price spike the release of public food stocks did occur in many countries, and proved a strong mitigating influence on the Food Price Crisis (FPC) in the developing world. The re-emergence of interest in physical food stocks has manifested itself, for example, in the creation of regional food reserves, in the role that food reserves have played in Brazil's successful *Fome Zero* campaign to eradicate hunger, and the reluctance of India to move away from physical food stock management as a key instrument of social protection (see Section 5.5.1).

The spike in world food prices which initially came in 2008 led to increasing food insecurity for millions, and associated riots in more than 30 countries. The subsequent drought in 2011 that hit the Sahelian region was the worst there in 60 years, turning a precarious situation into a crisis, which in Somalia resulted in a famine.

FAO has estimated that agricultural production needs to increase by 60% between 2005 and 2050 to enable the estimated nine to 10 billion people then to be adequately fed. USAID calculates that the world will need to do even more, namely to double its current food production, while at the same time climate change will be bringing increased droughts to many tropical regions, reducing their productive capacity. A report by ActionAid USA (2013), based on a review of recent economic modeling studies by researchers at Tufts University's Global Development and Environment Institute, finds that belief in a doubling of food production by 2050 to feed the extra billions are over-optimistic. Certainly, feeding the world for the next 35 years will be a challenge indeed, and the more so during the 35 years after that.

## REFERENCES

ActionAid USA, 2013. Rising to the challenge: changing course to feed the world in 2050 (Wise, T.A., Sundell, K.), 28 pp. <http://www.actionaidusa.org/publications/feeding-world-2050>.

Caulfield, L.E., Richard, S.A., Rivera, J.A., Musgrove, P., Black, R.E., 2006. Stunting, wasting, and micronutrient deficiency disorders. In: Jamison, D.T., et al., (Eds.), Disease Control Priorities in Developing Countries, second ed. World Bank, Washington, DC (Chapter 28).

Conway, G.R., 2012. One Billion Hungry: Can We Feed the World? Cornell University Press, Ithaca, NY, USA.

EC, 2009. Food Security: Understanding the Meaning and Meeting the Challenge of Poverty. EuropeanAid Cooperation Office, Brussels, 28 pp., p. 8.

Emmerij, L., Jolly, R., Weiss, T.G., 2001. Ahead of the Curve? UN Ideas and Global Challenges. Indiana University Press, Bloomington.

FAO, IFAD and WFP, 2014. The State of Food Insecurity in the World 2014: Strengthening the Enabling Environment for Food Security and Nutrition. FAO, Rome.

FAO Statistics Division, 2014. The State of Food Insecurity in the World 2013: An Overview. In: Asia and Pacific Commission on Agricultural Statistics. 25th Session, Agenda Item 9, Vientiane, Lao PDR, February 18–21, 2014, 5 pp.

HLPE, 2012. Food security and climate change. A Report by the High Level Panel of Experts on Food Security and Nutrition of the Committee on World Food Security, Rome. HLPE Report 3, June 2012. <www.fao.org/3/a-me421e.pdf>.

IDS, 2014. Gender and food security: towards gender-just food and nutrition security. Bridge Development—Gender. Institute of Development Studies, Sussex, UK, 104 pp.

Jolly, R., Emmerij, L., Ghai, D., Lapeyre, F., 2004. UN Contributions to Development Thinking and Practice. Indiana University Press, Bloomington.

Kracht, U., Schulz, M. (Eds.), 1999. Food Security and Nutrition: The Global Challenge. St Martin's Press, New York, NY.

Maxwell, D.G., 1996. Measuring food insecurity: the frequency and severity of coping strategies. Food Policy. 21 (3), 291–303.

Maxwell, S., Slater, R., 2003. Food policy old and new. Dev. Policy Rev. 21 (5–6), 531–553.

McDonald, B.L., 2010. Food Security. Polity Press, Cambridge, UK, 200 pp.

Psaki, S., et al., 2012. Household food access and child malnutrition: results from the eight-country MAL-ED study. Popul. Health Metr. 10 (24), published online December 13, 2012 http://dx.doi.org/10.1186/1478-7954-10-24.

Sen, A.K., 1981. Poverty and Famines; An Essay on Entitlement and Deprivation. Clarendon Press, Oxford, 257 pp.

Shaw, D.J., 2007. World Food Security—A History Since 1945. Palgrave Macmillan, Hampshire, UK, 472 pp.

Smith, M., Pointing, J., Maxwell, S., 1993. Household Food Security: Concepts and Definitions—An Annotated Bibliography. Institute of Development Studies, University of Sussex, 58 pp.

UNICEF, 2009. Tracking Progress on Child and Maternal Nutrition: A Survival and Development Priority. UNICEF, New York, NY, 124 pp.

Uvin, P., 1994. The International Organization of Hunger. Kegan Paul International, London.

# Chapter | Two

# Manifestations and Measurement of Food Insecurity

## 2.1 CONCEPTS OF HUNGER, UNDERNUTRITION, FOOD SECURITY AND NUTRITION SECURITY

### 2.1.1 Hunger

Hunger is a feeling of discomfort or weakness caused by lack of food, coupled with the desire to eat. It is used in this book as a close equivalent of chronic undernourishment (see below). In the first line of its website, WFP[1] points out that hunger is the world's number one health risk, killing more people each year than malaria, HIV/AIDS and tuberculosis combined. In 2014, FAO estimates indicate that global hunger reduction continues: about 805 million people were estimated to be chronically undernourished in 2012–14, down more than 100 million over the previous decade, and 209 million lower than in 1990–92. In the same period, the prevalence of undernourishment had fallen from 18.7% to 11.3% globally and from 23.4% to 13.5% for developing countries (FAO, IFAD and WFP, 2014) *op. cit.* The vast majority (98%) of hungry people live in developing countries, where almost 15% of the population is undernourished. Two-thirds of the world's hungry (526 million) live in Asia, while in sub-Saharan Africa, one in four people remains hungry. Latin America and the Caribbean is the region that has shown the greatest progress in hunger reduction, with the prevalence of hunger reduced by almost two-thirds since the early 1990s.

In the developing world, in addition to the millions of chronically undernourished, another 5–10% are at risk of acute food insecurity in times of crisis. Despite improvements, the Millennium Development Goal (MDG 1) on eradicating extreme poverty and hunger was likely to be missed by a wide margin in sub-Saharan Africa, especially where persistent food insecurity is

[1] Retrieved from www.wfp.org/hunger/stats (accessed June 3, 2015).



Food Security in the Developing World. DOI: http://dx.doi.org/10.1016/B978-0-12-801594-0.00002-6

compounded by widespread political instability, conflict and the HIV/AIDS pandemic. Countries that have experienced such in recent decades are more likely to have sustained significant setbacks in reducing hunger. Landlocked countries face persistent challenges in accessing world markets, while countries with poor infrastructure and weak institutions also face constraints (Case Study 8 on the book's companion website, http://booksite.elsevier.com/9780128015940).

Humans need to take in an appropriate amount of food to satisfy their emerging hunger, thereby fulfilling their nutritional requirements and maintaining normal metabolism. The amount needed depends on factors like age and degree of activity, and the proportion of the ingested nutrients which are absorbed into body tissues. A "balanced" diet is needed for the physiological requirement to be satisfied, though there can be a mismatch between quantity and quality of intake. *This may lead to hunger being satisfied, but not all the nutritional needs.* In this case, the body exhibits a degree of dysfunction, manifested in various ways such as inertia, and specific morphological and/or physiological nutritional deficiency symptoms. Deficiencies experienced during childhood can be irreversible, and lead to adults who are abnormal in terms of height and/or weight for age, and/or having impaired cognitive function.

An article in the medical journal *The Lancet* in 2008 calls for hunger to be prioritized, to match the attention given to other issues such as climate change (Sheeran, 2008). The "Zero Hunger Challenge" (ZHC) is a global call-to-action which aims to build support around the vision of achieving Zero Hunger. It was launched in 2012 by the UN Secretary General and calls on every stakeholder—governments, the private sector, NGOs and the public—to do its part to turn this vision into a reality. Based on a shared conviction that hunger can be eliminated in our lifetimes, the vision has five key elements:

1. Zero stunted children of under 2 years of age
2. 100% access to adequate food all year round
3. All food systems are sustainable
4. 100% increase in smallholder productivity and income
5. Zero loss or waste of food

The issue of a "Hunger Index" is raised in Case Study 6, companion website, in relation to food insecurity in Lao.

### 2.1.2 Undernutrition

Humans need nutrients to enable them to metabolize, grow and develop. A mix of energy-giving and body-building ingredients, together with water as the solvent, is needed. An individual can remain healthy if these nutrients are provided in a sufficient and regular way, are adequately ingested and absorbed, and translocated to the tissues where they are needed.

If, through inadequate or imbalanced intake and/or poor absorption and/or poor biological use of nutrients consumed as a result of repeated infectious disease, one or more of these nutrients is/are not continually availed in this way, up to the level of physiological need, then the individual will become undernourished. Often the undernourished condition is pronounced, as seen in the aftermath of many disaster situations. However, the onset of undernutrition may be less pronounced, though the individual may still be chronically undernourished. Over time, undernutrition is associated with morbidity and even death, infants and young children being among those most seriously impacted. Morbidity brings with it reduced educational and work outcomes, and often poverty, which positively reinforces the undernutrition.

Though the term undernutrition is often used in the more restricted way, referring to an individual unable to acquire enough food to satisfy dietary *energy* requirements (only), in developing countries there is a range of nutritional deficiency diseases of humans, as discussed below, some of which are uncommon in developed countries. As discussed further in chapter 3 "Causes of Food Insecurity", undernutrition can arise in one or more ways, for example:

- conflict situations in which food supply and quality is compromised
- situations in which a natural disaster has struck and destroyed the food stocks in storage or at retailers, or the standing crops washed away, desiccated through drought, or destroyed by wild animals/pests/disease
- a dysfunctional socio-economic environment in which food is available but an individual or community cannot access it, due to poverty or physical/social access to it denied
- the food available is unsafe to eat, due to spoilage or contamination
- the food available or consumed is nutritionally poor, through suboptimal knowledge or feeding practice, sometimes related to prevailing social norms.

Hunger and undernutrition kill nearly 2.6 million children per year. Undernutrition affects one in three children in developing countries. Often the problem starts before birth due to undernourishment of the mother. Undernutrition is a root cause of vulnerability, especially for children up to the age of 2. It lowers intellectual and physical development, thereby reducing the capacity of tomorrow's adults to cope with adverse events. Moreover, it costs many developing countries up to 2–3% of their Gross Domestic Product (GDP) each year, extending the cycle of poverty and impeding economic growth.

### 2.1.3 Food Insecurity and Nutrition Insecurity

The two terms are associated, but describe different conditions, each having a number of direct and indirect causes. Food insecurity is a quantitative term, whereas nutrition insecurity is more qualitative in nature. Food security does not ensure nutrition security. Kaufmann (2008) documented populations in The People's Democratic Republic (PDR) of Lao (Case Study 6, companion website), which were nutritionally insecure despite having plenty of food to hand and not being poor. She documented the various social malaises responsible, these largely rooted in poor healthcare in the widest sense, and norms which undervalued women—the mothers of the children who were also nutritionally insecure.

As a second example of food and nutrition security being uncoupled, at a Community Management for Acute Malnutrition (CMAM) clinic in northern Nigeria in the late 2014, the current author witnessed young mothers lining up with their undernourished infants waiting for Ready-to-Use Therapeutic Food (RUTF) to be dispensed, mothers who were themselves fairly well nourished, beautifully made-up and in fine quality dresses, clearly not poor. Insufficient awareness of what constituted nutritious diets for their infants, together with certain social norms pertaining to child feeding practices, was the direct cause of the latter's stunted condition, due to insufficient protein, minerals and vitamins.

Food security is a necessary, but insufficient, condition underlying *comprehensive* nutrition security. However, even in the presence of food insecurity and poverty, the status of nutrition security can be at least partially improved through changes in food-related behavior patterns, improved literacy and knowledge on diet for child caregivers (the birth mother or other person), or food fortification which improves the status of a micronutrient in the diet. Kaufmann (2008) *ibid.* instances such in Lao PDR.

The commonly accepted World Food Summit definition of food security (see Section 1.1) implies that nutrition security will automatically follow from food security. It *assumes* that households which have economic and physical access to good quality food which is available near to the home, will buy wisely, prepare and store food wisely, ensure that adequate amounts are consumed in a regular way and that sufficient benefit is obtained from ingesting that food. However, one or more of these suppositions may not pertain for a given household or individual in it. That beneficial nutritional outcome is not stated more robustly in the standard definition of food security represents a weakness, an insufficient statement that through its brevity omits some determining factors which can prevent food security being consummated by nutritional security. Practitioners in the field seeking to improve the nutritional status of individuals and communities understand that even if food availability and economic and physical access to it is assured, there may be social and cultural norms which preclude adequate use of, and benefit from, that food.

Nutrition insecurity will likely have a negative feedback on food security, through reducing the capacity of an individual to obtain adequate nutritious food, and derive the needful benefit from it. Dual causality applies, nutritionally unfilled individuals being more likely to become food-insecure, and food-insecure people more likely to become nutritionally-insecure. The vicious circle can have many facets; for instance, where infectious disease is a determinant of undernutrition, provision of more food will not much help. Conversely, if insufficient food is available or accessed, provision of better healthcare will be of limited value.

Those involved in planning or implementing measures to relieve food insecurity often concentrate on promoting food production/availability, with less attention given to nutritional security. As a specific instance, this was pointed out in the final external evaluation of the EU Food Facility program following the 2008 food price crisis, and the European Union has taken measures to address that in its future programing. This book endeavors to give voice to both food security and nutrition security.

Nutrition security requires simultaneous availability and access to safe food, good health and good care in general (including hygiene, clean drinking water and proper feeding of a young child who is not old enough to feed independently of the caregiver).

At the UN Standing Committee on Nutrition "Meeting of the Minds" session from March 25–28, 2013, the following definition was proposed to cover both food and nutrition security "Food and nutrition security exists when all people at all times have physical, social and economic access to food, which is consumed in sufficient quantity and quality to meet their dietary needs and food preferences, and is supported by an environment of adequate sanitation, health services and care, allowing for a healthy and active life". The absence of these conditions (together with the uncited "clean drinking water") results in food and nutrition insecurity. The reader is referred to FAO (2009) for a further comparative discussion on food and nutrition security.

## 2.2 NUTRIENT DEFICIENCY DISEASES

*Undernutrition* is an umbrella term and has been classified into a number of forms according to symptoms which the body exhibits, overtly with some conditions though less so with others, which may even be "sub-clinical" and need special tests to determine. Each represents an instance of sub-optimal growth and/or development, such impairment represented by three basic dysfunctional conditions—protein-calorie, mineral and vitamin.

### 2.2.1 Protein-Calorie Undernutrition

*Acute undernutrition* refers to a "wasting" condition, a loss of body weight. This condition when at its most pronounced is called *marasmus*. A child suffering from this condition is both food- and nutritionally-insecure, and will not live long unless its condition is rapidly addressed. Skeletal features become more pronounced, such as the facial bones, ribs and knee joints, the body diminishing to become "all skin and bones". Locomotion and movement become labored as muscle mass decreases and loses tone, until eventually movement all but ceases, with negligible muscle volume or energy whereby they can function. The individual may remain alert however. The cause of the condition is insufficient intake and/or absorption of nourishment in general, such that demand exceeds supply and the body metabolizes its own tissues by default. Famines in Ethiopia, the most serious for a century in 1984–85 for instance, brought television images of such emaciated children into the living rooms of many better-off people around the world. Recovery from this condition through provision of food is a slow process. If the condition is advanced, nourishment needs to be given intravenously or via a tube through the nose directly into the stomach.

*Severe acute undernutrition* refers to a condition in which body mass may not be less than "normal", but much of that mass is tissue fluid. The body appears obese, due to fluid retention (edema), which is especially noted in the lower limbs and face. A thumb pressed onto the skin surface causes a depression which remains for a while after the thumb is removed. In severe cases, the skin becomes ulcerous, the hair color lightens and the hairline recedes. Unlike marasmus, the individual has a dull and withdrawn demeanor, socially disengaged and does not smile or interact with others. This condition is called *kwashiorkor*, and can be mild to severe, and largely affects children. The cause of the condition is an imbalance of food intake, with insufficient protein compared with carbohydrate and/or oil/fats. A child suffering from this condition is food-secure in the sense that it is not hungry, but nutritionally highly insecure, and may well die unless more protein is incorporated into its diet. Once this happens, there is a fairly rapid disappearance of the external symptoms, one of the first stages of which being an initial loss of weight as the edema dissipates.

*Chronic undernutrition* is a condition of "stunting" (shortness) compared with the normal height for a given age. This condition develops over a longer time-frame than acute undernutrition, from which it is thereby distinguished. There are many data showing that the period in a child's life which is most vulnerable to chronic undernutrition developing is the first 18 months. Poor nutrition in the pregnancy of the mother pre-disposes the fetus to damage and retardation even before birth. Nutritional interventions to improve mother and child are best made over this 27-month period, when they will have the greatest impact.

#### 2.2.1.1 OBJECTIVE MEASUREMENTS OF PROTEIN-CALORIE UNDERNUTRITION

The most frequent means of assessing whether protein-calorie undernutrition is present are through non-invasive body measurements (anthropometry). This involves:

- weighing a child and measuring the child's height, and comparing the measured weight against a WHO reference weight for the height of the individual in the population being studied. If the measured weight falls below the norm, then the child is deemed to be undernourished. In statistical terms, if the degree by which the weight-for-height measurement falls short of the WHO standard norm is less than two "standard deviations" then the child is considered to be 'wasted'; if the shortfall is three standard deviations then the child is considered to be suffering from severe acute malnutrition
- measuring the child's mid-upper arm (bicep/tricep) circumference (MUAC); a reading of less than 12.5 cm indicates acute undernutrition (wasting); a reading of less than 11.5 cm indicates severe acute undernutrition

**TABLE 2.1** Wasting and Stunting by Country Among Children 12–23 Months[3]

| | | Wasting (<−2 WHZ) (tercile) | | |
|---|---|---|---|---|
| | | Low | Mid | High |
| Stunting <−2 HAZ (tercile) | Low | Armenia; Bolivia; Colombia; Dominican Rep; El Salvador; Jordan; Moldova; Swaziland | Azerbaijan, Ghana | Maldives |
| | Mid | Honduras | Congo (Brazzaville); Egypt; Kenya; Namibia; Sierra Leone; Uganda; Zimbabwe | Bangladesh, Guinea, Haiti, Mali |
| | High | Guatemala, Zambia | Benin, Cambodia | DRC (Congo Kinshasa); Ethiopia; India; Nepal; Nicaragua; Niger; Nigeria |

HAZ, height for age Z-score; WHZ = weight-for-height Z-score.
*Source: Reproduced from Bergeron and Castleman (2012) (with permission).*

- assessing edema in the lower leg tissue or feet, by applying steady digital pressure and noting if the depression remains for a while after the assessor's hand is removed. If it does, then the child is suffering from kwashiorkor. This and the arm circumference measure, provide a rapid assessment tool in humanitarian emergency situations.

Measuring undernutrition in emergency situations is well-dealt with by UNICEF.[2]

#### 2.2.1.2 DISTRIBUTION OF WASTING AND STUNTING CONDITIONS

"Wasting" and "stunting" generally coexist within populations. Table 2.1 shows how 34 countries from across the world feature in this respect.

An article in *The Lancet* of 2008 (one of a highly regarded series on undernutrition in the journal that year[4]) shows that 80% of the world's stunted children live in just 20 countries, and that intensified nutrition action in these countries

---

[2] www.unicef.org/nutrition/training/list.htm.

[3] This table was created with MEASURE DHS StatCompiler, available at http://www.measuredhs.com/data/STATcompiler.cfm. All countries with the data available to classify the country were used. This resulted in the inclusion of the following surveys: Bangladesh 2007, Benin 2006, Bolivia 2008, Burkina Faso 2003, Burundi 1987, Cambodia 2005, Cameroon 2004, Chad 2004, Congo 2005, Cote d'Ivoire 1998–99, DRC 2007, Ecuador 2004, Eritrea 2002, Ethiopia 2005, Gabon 2000, Ghana 2008, Guatemala 2008, Guinea 2005, Haiti 2005–06, Honduras 2005–06, India 2005–06, Kenya 2008–09, Lesotho 2004, Malawi 2004, Mali 2006, Mauritania 2000–01, Morocco 2003–04, Mozambique 2003, Namibia 2006–07, Nepal 2006, Nigeria 2008, Peru 2000, Sierra Leone 2008, Swaziland 2006–07, Tanzania 2004–05, Uganda 2006, Yemen 1997, Zambia 2007, Zimbabwe 2005–06.

[4] www.thelancet.com/series/maternal-and-child-undernutrition.

could lead to achievement of MDG 1c, and greatly increase the chances of achieving goals for child and maternal mortality (MDGs 4 and 5) (Bryce et al., 2008). However, development efforts in most countries having high rates of undernutrition fail to reach undernourished mothers and children with effective interventions, supported by appropriate economic and social policies.

Another key paper on nutrition, using statistics on stunting of children in the first 5 years of their life, was published by a team from the World Health Organization (WHO), Geneva (De Onis et al., 2012). The paper collated 576 national surveys in 148 developing and developed countries. In 2010, it is estimated that 171 million children (of which 167 million in developing countries) were stunted. Globally, childhood stunting decreased from 39.7% in 1990 to 26.7% in 2010. This trend is expected to reach 21.8% or 142 million in 2020. While in Africa, stunting has stagnated at about 40%, Asia showed a dramatic decrease from 49% in 1990 to 28% in 2010, the number of stunted children nearly halving from 190 million to 100 million. This trend in Asia is expected to continue, so that in 2020, Asia and Africa will likely have similar numbers of stunted children (68 and 64 million, respectively). Rates are much lower (14% or 7 million in 2010) in Latin America.

### 2.2.2 Micronutrient Deficiencies

In addition to the protein-calorie deficiencies mentioned in Section 2.2.1, there are micronutrient deficiency conditions, caused by shortfall of one or more minerals or vitamins in the consumed food. Such micronutrient deficiencies are normally correlated with overall undernutrition, with a few exceptions such as iodine. Millions of people are nutritionally insecure in this way. In 1997, this condition became graphically branded by UNICEF as the "silent urgency".

Minerals and vitamins (derived from the term "vital amines") are essential for good nutrition, especially iron, iodine, zinc, vitamin A, folate (vitamin B9), thiamine (vitamin B1), riboflavin (vitamin B2), niacin (vitamin B3), vitamin B12 and vitamin B6, vitamin C and D. Micronutrients are required for many metabolic processes in the human body, yet only in *small quantities*. The physiological and development conditions associated with micronutrient deficiencies in the diet affect the health and lives of up to 2 billion people, in both developing and developed countries, regardless of age and gender, though with some groups being at particularly high risk.

*Each year more than 1 million children under 5 years of age die of Vitamin A and zinc deficiencies (*Micronutrient Initiative et al., 2009*)*. Not only do micronutrient deficiencies have a direct negative impact on human health, but they also increase the severity of infectious diseases such as measles, tuberculosis and HIV/AIDS, thereby representing both a huge public health challenge and opportunity (Tulchinsky, 2010). Micronutrient deficiencies are not always clinically apparent, and may relate not to food availability or access, but to place of residence, or religious or dietary practices. Some of the key mineral and vitamin deficiency conditions are mentioned in the following sections.

#### 2.2.2.1 MINERAL DEFICIENCIES

##### 2.2.2.1.1 Iron

Iron is found in all plant foods but is more plentiful and bioavailable in meat. Deficiency results from insufficient absorption of iron or excess loss. When there is insufficient iron in the body, fewer red blood cells are produced, reducing the capacity of the blood to transport oxygen. *Iron deficiency anemia is the most prevalent nutritional disorder in the world*, and not only in the developing world. More than 2 billion people, a third of the world's population, are anemic, many due to iron deficiency.[5] Iron deficiency anemia is most prevalent in South Asia and sub-Saharan Africa.

Major health consequences include fatigue and inability to concentrate, reduced resistance to infections, poor pregnancy outcome, impaired physical and cognitive development and increased morbidity in children. Iron deficiency constitutes a public health condition of epidemic proportions, particularly for women of reproductive age and children. It exacts a huge toll in terms of ill-health, reduced school performance, premature death and lost productivity and earnings, invisibly eroding the development potential of individuals, communities and national economies. The Micronutrient Initiative et al. (2009) *ibid.* estimates that more than 130,000 women and children die each year because of iron deficiency anemia.

WHO (2014) considers that the most vulnerable are the poorest and the least educated, and who therefore stand to benefit most by the condition's redress. The solution is both inexpensive and effective, through dietary diversification with iron-rich foods, food fortification and iron supplementation. An important corollary counter-measure is better control programs for diseases which aggravate iron deficiency, such as malaria and tuberculosis, and internal parasites such as hookworm and the *Schistosoma* worm which is responsible for Bilharzia.

##### 2.2.2.1.2 Iodine

Iodine is essential for human development and growth, exerting its influence through the endocrine system, being essential for the formation of the thyroid hormones which regulate growth and development. Deficiency can lead to fetal loss, congenital abnormalities and hearing impairment. Severe deficiency in adults leads to the visual symptom of goiter, a swelling of the thyroid gland in the neck. Deficiencies while young can lead to cretinism, with its attendant mental retardation and associated physical disabilities. The majority of iodine-deficient individuals, however, experience only mild mental retardation, which nevertheless has significant economic consequences. It is estimated that 2 billion persons worldwide consume insufficient iodine. A high prevalence of iodine deficiency disorders occurs in eastern Europe and central Asia, the eastern Mediterranean and north Africa, south Asia and

---

[5] Though iron deficiency is the prime cause of anemia, the condition can also result from Vitamin A, C or folate deficiency, malaria or HIV.

sub-Saharan Africa (Zimmerman, 2009). Iodized salt is a simple remedy, leading to swelling in the neck subsiding.

##### 2.2.2.1.3 Zinc

This mineral is an essential part of many enzymes and plays an important role in protein synthesis, and cell division and differentiation. It regulates many bodily functions, including tissue growth and healing of wounds, proper thyroid function and clotting of blood, immune function, metabolism of proteins, fats and carbohydrates, cognitive ability, fetal development and sperm production. The symptoms of severe deficiency include retarded growth (stunting), diarrhea, mental disturbances and recurrent infections (WHO and FAO, 2006). The geographical regions most affected include South Asia (Bangladesh and India in particular), Africa and the western Pacific. *Zinc deficiency is estimated to be responsible for about 800,000 deaths annually*, through diarrhea and low resilience against pneumonia and malaria in children under 5 years of age, especially in sub-Saharan Africa, eastern Mediterranean and South Asia. Only animal flesh (particularly shellfish) is a good source of zinc, so that like iron deficiency, populations having a plant-based diet are susceptible to zinc deficiency.

#### 2.2.2.2 VITAMIN DEFICIENCIES

##### 2.2.2.2.1 Vitamin A Deficiency

Vitamin A is essential for the body's immune system to help fight infections, and also for proper vision, growth and reproduction. Vitamin A deficiency (VAD) is the leading cause of blindness in children, with up to half a million children becoming blind each year for this reason. According to WHO and FAO (2006) *ibid.*, half of these children die within a year of becoming blind. Vitamin A deficiency also causes night blindness and xerosis of the eye cornea and conjunctiva, and increases the risk of maternal deaths and child deaths especially from diarrhea and measles. Deficiency also increases the risk of death during pregnancy for both mother and fetus, and after birth for the newborn. VAD is a **major public health problem in the developing world** affecting 190 million children under 5 years of age, particularly in Africa and South East Asia. Dietary sources of pre-formed Vitamin A include liver, milk and egg yolk. Dark green leafy vegetables, and yellow and orange non-citrus fruit and vegetables are common sources of Vitamin A precursors which can be converted to Vitamin A in the body; these are generally less bioavailable than from animal sources but tend to be more affordable.

##### 2.2.2.2.2 Vitamin B Deficiencies

Vitamin B deficiencies are highly prevalent in many developing countries, especially where diets are low in animal products, fruit and vegetables, and where cereals are milled prior to consumption. Pregnant and lactating women, infants and children are most at risk of Vitamin B deficiencies.

*Thiamine (Vitamin B1):* Severe thiamine deficiency can result in the condition "beri-beri", potentially fatal heart failure or peripheral neuropathy.

*Riboflavin (Vitamin B2):* Early symptoms of deficiency can include weakness, fatigue, mouth pain, burning eyes and itching. More advanced deficiency can cause brain dysfunction.

*Niacin (Vitamin B3):* Niacin deficiency can result in "pellagra", skin rashes being a symptom, together with vomiting, diarrhea, depression, fatigue and loss of memory.

*Pyridoxine (Vitamin B6):* Symptoms of severe deficiency include neurological disorders (epileptic convulsions), skin changes and possibly anemia.

*Folate (Vitamin B9):* Folate plays a key role in cell multiplication and tissue growth. Deficiency leads to the risk of neural tube defects, a condition that affects an estimated 300,000 newborns each year (WHO & FAO, 2006) *ibid.* Folate deficiency can also lead to impaired cognitive function in adults. This deficiency condition is often associated with populations which consume in their diet a lot of cereals, which are low in folate, and few leafy greens and fruit, which are rich in it.

*Cobalamin (Vitamin B12):* Deficiency of Vitamin B12 causes neurological deterioration, megaloblastic anemia and possible impaired immune function. Deficiency can severely delay the development of infants and young children.

#### 2.2.2.3 Vitamin C

Deficiency leads to scurvy, associated with fatigue, hemorrhages, low resistance to infection and anemia. The condition was first understood through sickness and death of sailors, and the condition corrected by their being supplied with fruits such as lime during their sea voyages. Green vegetables are another rich source of this vitamin.

#### 2.2.2.4 Vitamin D

Deficiency leads to rickets, osteomalacia, osteoporosis leasing to bone fractures, cardiovascular conditions and colorectal cancer. Deficiencies are widespread across all age groups and are associated with low exposure to sunlight, particularly for dark-skinned people living in northern latitudes and among youngsters spending a lot of time in front of computers instead of outdoors, the elderly and others who tend not to live much in the open.

Of the micronutrient deficiency conditions listed in Section 2.2.2, WHO (2000) regards iron, iodine, Vitamin A and zinc deficiencies as having the most serious global significance.

## 2.3 NUTRIENT OVERSUFFICIENCY

Drawing partly on study data published by Stevens et al. (2012), which was the first report of adult overweight and obesity prevalence by country, year and sex, an ODI report by Keats and Wiggins (2014) showed that the number and proportion of overweight and obese adults in both developed and developing countries have greatly increased over the last few decades. According to the World Health Organization (WHO), there will be in 2015 about 2.3 billion overweight people aged 15 years and above, and over 700 million obese people worldwide.

Globally, the percentage of adults who are classified as overweight or obese (having a body mass index (BMI) greater than 25), grew from 23% to 34% (1.46 billion) between 1980 and 2008, the majority of these additional cases being in developing countries. The greatest increase in incidence of overweight people occurred in South East Asia, in which the percentage more than trebled from a low starting point of 7% to 22%; in China and Mexico overweight and obesity occurrence almost doubled and in South Africa it had increased by a third. A total of 904 million people in developing countries have a BMI of more than 25, up from just 250 million in 2008, an increase of 3.6 times; in developed countries some 557 million fall into this category, having increased 1.7 times over the same period. The report predicts an associated increase in the incidence of non-communicable diseases—diabetes, heart attacks, strokes and some cancers, all of which add to the burden on public health facilities and budgets. Over the 28-year period reported on, the global population almost doubled.

The rise in obesity is happening in parallel with the protein-calorie food and micronutrient undernourishment crisis in the developing world as cited in Section 2.2 above, affecting hundreds of millions, particularly children. Obesity and undernutrition can and do exist side by side, sometimes in the same household. *It is in the developing world that there is both the greatest acceleration in food overconsumption and the greatest continuing iniquity of underconsumption.*

In obesity-affected developing countries, incomes are rising and there has been a dietary shift from cereals to more fats and oils, sugar and animal products. A more sedentary lifestyle, the increased availability of processed foods and influence of media advertising have been reasons for the change in diet. To correct this deterioration of nutrition-related health, the ODI Report recommends more concerted public health measures by governments, as successfully done in South Korea, for example, in terms of policy and awareness campaigns, akin to measures taken in the developed world to limit smoking.

## 2.4 CHALLENGES OF ACCESSING AND INTERPRETING FOOD SECURITY AND NUTRITION SURVEY DATA

### 2.4.1 Challenges in Measuring Household Food Security

Most assessments of household food security rely on measuring food consumption. Two major methods have been widely used. The first (used by economists) is to estimate gross household production and purchases over a period of time, and also the growth or depletion of food stocks held over that period of time (assuming that the food which has come into the household's possession and "disappeared" has been consumed). The second method (used by nutritionists) is to undertake 24-h recalls of food consumption for individual members of a household, and assess each type of food mentioned for calorific content. This latter method results in more reliable consumption data and captures intra-household distributional differences which the first method does not measure. Both methods mostly capture only the sufficiency element, rather than sustainable economic access or vulnerability, so neither method has been accepted as optimal to assess household food security.

Alternative *proxy* indicators used to estimate or predict food security at household level have included food balance sheets, rainfall and marketing data, asset ownership, household size and dependency ratio. Another approach has been to examine the use of, and reliance upon, a *direct* measure, namely household coping strategies for dealing with insufficiency of food (Maxwell, 1996) (see Section 4.1).

Psaki et al. (2012) *op. cit.* conducted a study across eight countries (rural, urban and peri-urban areas of Bangladesh, India, Nepal, Pakistan, Peru, Brazil, South Africa and Tanzania) to investigate the relationship between household food *access* (one component of food security), and indicators of nutritional status in early childhood across eight country sites. The study thereby addressed a common concern that conclusions from studies are location-specific. Psaki's study provided evidence of the validity of using a simple household food access insecurity score to investigate the etiology of childhood growth faltering across diverse geographic settings. Such a measure could be used to direct interventions by identifying children at risk of illness and death related to undernutrition.

Food security is but one element of livelihood security, and indicators of the former should not be interpreted independently of a good understanding of the latter (Chambers, 1988; Maxwell and Smith, 1992; Davies, 1993). Frankenberger and Coyle (1993) have observed that "poor people balance competing needs for asset preservation, income generation and present- and future-food supplies in complex ways, and may go hungry up to a point to meet other objectives". Inferring food security *solely* from consumption data is unsafe.

### 2.4.2 Challenges in Measuring Undernutrition

A study in Yemen illustrates some of the challenges facing nutrition professionals in developing their programs at country level. In April 2014, UNICEF put out a Call for Expressions of Interest from research institutions which could partner it to conduct an undernutrition causality study in Yemen, to be run from the UNICEF country office. The Call made for interesting reading, for example regarding the challenge of comparing existing data sets, like for like. Yemen, which ranks 154 out of 187 in the UNDP Human Development Report of 2011, is categorized as suffering from serious undernutrition, according to the WHO (1995) thresholds of problem severity. Chapter 12 of the Yemen government Ministry of Health Family Health Survey Report of 2003 shows a dramatic increase in undernutrition compared with the level identified by the earlier USAID-funded Demographic and Health Surveys (DHS) for Yemen of 1991–92. However, because nutrition status was not included in the earlier survey report, UNICEF needed to extract the missing data set files from the DHS website (www.measuredhs.com). Moreover, figures in both the 1991–92 and 2003 surveys for stunting and wasting criteria did not match the WHO Growth Standards, so the figures needed transforming in each case.

Another challenge for UNICEF in identifying the severity of undernutrition deterioration in Yemen since 1991, is that the DHS survey then and the five subsequent surveys by various agencies dealt with *some* of the determinants of nutritional status as laid down in the UNICEF conceptual frame work, but none considered *all* of them ! Moreover, in UNICEF's view, the quality of some tools used for data collection, and the anthropometrical data itself are questionable. SMART (Standardized Monitoring and Assessment of Relief and Transitions) surveys conducted in 2012 and 2013 did not provide many data on determinants of nutrition status, though they did serve to identify "hotspots" of localized undernutrition (NB. SMART survey methodology is based on the two most vital and basic public health indicators for the assessment of the magnitude and severity of a humanitarian crisis—nutritional status of children under 5 years of age and mortality rate of the population).

As a result of all this, UNICEF saw fit to put out a Call for Expressions of Interest to Universities and Research Centers, so that a comprehensive survey might be undertaken to assess the current status of undernutrition in Yemen. The data would inform an action plan to best address undernutrition in Yemen, and also set a benchmark baseline study data set against which future UNICEF survey results could be assessed.

## 2.5 DIFFERENTIAL VULNERABILITY TO FOOD AND NUTRITION INSECURITY

### 2.5.1 Communities

#### 2.5.1.1 RURAL POOR

Poor people are highly vulnerable to food and nutrition insecurity, and these tend to be smallholders, marginal pastoralists and people in urban slums. Despite the burgeoning of urban slums over recent decades, nearly 75% of poor people in developing countries live in rural areas[6]. Most of the 795 million hungry people in the world are smallholder farmers who depend on agriculture for their livelihoods and to feed their families, and most of these

[6] In South America, the opposite tends to be the case, with the poor concentrated in urban slums (see Case Studies 4 and 5 on the book's companion website).

farmers are women. USAID (2014)[7] points out that growth in the agricultural sector is twice as effective as growth in other sectors in reducing poverty.

#### 2.5.1.2 SLUM DWELLERS

Undernutrition is sometimes wrongly believed to be just a rural phenomenon, with many surveys showing a higher incidence of undernutrition in urban areas. While working in Kebbi State in northern Nigeria in 2014, the current author noted that the health clinic visited in the periphery of the capital city Birnin Kebbi had many stunted infants to whom the staff were attending. Despite Kebbi being a livestock-rich State with a good supply of milk in town, the caregivers could either not afford to buy that milk and/or were following the customary local practice of giving their infants water rather than cow's milk when taken off the breast.

Case Study 10 on the book's companion website gives an overview of the dire living conditions in urban slums. The nutritional status of slum children is the worst of all urban groups, and worse than the rural average. Urban drifters, who take up residence in slums of financial necessity, find no relief from the poverty and undernutrition they knew earlier in the countryside. In most global slum populations, a distressing feature is the lack of any significant improvement over the years.

Cities in India provide an example where urban slums are rampant and undernutrition rife. Urban slums in India have problems quite unlike the country's rural areas (Jayarajan, 2014). The health infrastructure in cities is managed by urban local bodies (municipal corporations and development authorities) yet they do not take full account of the real issues of slum residents. Slums are categorized into declared and undeclared slums, the latter being worse off. Declared slums are those that have been recognized by the national government concerned as having poor living conditions, whereas undeclared slums have similar challenges, but do not receive support under any of the slum rehabilitation programs which are available to the former. The accessibility and provisions of government facilities and schemes are skewed towards the declared slums. So for instance, there are no *Anganwadi* Centers in undeclared slums—these Centers started in 1975 as part of the Integrated Child Development Services (ICDS) to combat child hunger and undernutrition. They provide government-sponsored child and mother care programs as part of the public healthcare system, catering for children from birth to 6 years old.

Spiraling food prices and death of children due to severe undernutrition were in the spotlight in India as 2010 drew to a close. Shockingly, undernutrition is almost never cited as a cause of death, but its impact among slum children in Mumbai (and other cities in Asia) is alarming (Rockefeller Foundation, 2011). Figures then available showed that Mumbai's slums housed around 730,000 children of ages 4 years old and below. Estimates

[7] Retrieved from www.usaid.gov/what-we-do/agriculture-and-food-security (accessed April 26, 2014).

indicated that nearly 40–60% of these slum children were undernourished, and of these, 7.5–30% were severely so. Indian government data show that about 3.5% of Mumbai's slum children die before they turn 6 years. Even those children who survive display severe school dropout rates, and a low ability to develop skills and break the poverty trap.

A study conducted by Goel et al. (2006) in the urban slums of Rohtak, a city in Haryana, on 540 children aged 1–6 years showed that 57.4% (310) of children were undernourished, despite these slums being served by the ICDS project. WHO and UNICEF in *The State of the World's Children 2014* give figures of 48% moderate and severe stunting, and 20% moderate and severe wasting, for India as a whole. Protein-energy insufficiencies or imbalance, anemia and vitamin A deficiency plague a large proportion of Indian children.

#### 2.5.1.3 URBAN COMMUNITIES IN CONFLICT

Civilian communities in specific locations, like those trapped in Homs town during the Syrian civil war which started in 2012, became food-insecure as a result of the siege laid by government forces. Homs was a rebel stronghold and the Assad regime deliberately targeted it both militarily and by cutting off food supplies to the town, which eventually forced the total withdrawal of the fighters by May 9, 2014. While under siege, the civilian population suffered, through both the fighting and lack of food. Refugees from Syria who managed to reach Jordan, for instance, are in receipt of food, water and other essentials from international organizations at camps constructed to house them within Jordan, until the available money runs out.

The United Nations Relief and Works Agency (UNRWA), which specifically targets Palestinian refugees in the Near East, frequently runs out of money because of the high demand on its resources. In 2009, for instance, UNRWA had totally run out of money to purchase food to feed Palestinian refugees, largely in the blockaded Gaza Strip, which is urban/peri-urban. The refugees there were in a dire condition, the Strip having just endured the horror of the Israeli military's 3-week offensive *Cast Lead* operation (December 2008 until January 2009). Some 1,387 people were killed, the majority being non-combatant civilians, including 320 children under 18 years of age (B'Tselem, 2009). The population of the Strip was already in dire straits because of the Israeli embargo on goods entering the official crossing points through the Separation Fence which the Israeli government first constructed in 1994. Fortuitously, at the time the Euro 1 billion EU Food Facility instrument to relieve the impact of high food prices for 49 of the most affected countries was under way, and FAO and other UN organizations (UNRWA, WFP, OCHA) spear-headed the case for Palestine to be a beneficiary of this. On March 30, 2009, Euro39.7m was approved for allocation to Palestine by the European Commission, all of which was assigned on May 26 to UNRWA for its social safety net program. This enabled UNRWA to purchase food at a time of record high world food prices, this saving the lives of countless Palestinians.

**FIGURE 2.1** Local staff of Oxfam GB in Gaza holding a vigil on March 14, 2013. With permission from Elena Qleibo, Oxfam GB Food Security and Livelihoods Coordinator, Gaza.

The enormous resilience to adversity shown by Gazans is legendary. Despite suffering themselves from being in what has been called the 'largest prison in the world', Gazans still made time to show solidarity with the strife-torn people of Syria in their time of need (Fig. 2.1), the picture uploaded onto Twitter as part of a global "storm" of support that day.

#### 2.5.1.4 SMALL ISLAND DEVELOPING STATES

Small Island Developing States (SIDS), of which there are 52 listed by the UN Department of Economic and Social Affairs, share many vulnerabilities to food insecurity. These include remoteness from major markets, and related high transport costs and susceptibility to external shocks which affect the imported food bill; small size so that economies of scale are difficult for agricultural and other industries; small but growing populations putting pressure on limited natural land and fresh water resources; and, susceptibility to natural disasters, including violent weather and rise in sea level. Two examples are given below—Jamaica in the Caribbean grouping of SIDS, and the Independent State of Samoa, in the Polynesian region of the Pacific Ocean[8].

Jamaica is relatively large, 90 mi south of Cuba and 445 mi from Columbia in South America. It has an area of 4,240 mi$^2$ (10,981 km$^2$), and a population of 2.778 million in 2015. By contrast, Samoa is much more remote from any mainland, 2,450 mi from Brisbane, and far smaller. It comprises two main islands, Upolu and Savai'i representing 99% of the land mass, together with three small islets. Its area is 1,097 mi$^2$ (2,841 km$^2$), with a population of some 200,000.

[8] The Central American country of Belize, with more than a thousand islands or "*cayes*", is classified as a SIDS, and discussed generically in Case Study 3 on the book's companion website.

A food security challenge for both Island States is the non-competitive local agricultural industry. Though there is an abundant supply of fish, fruit and vegetables, the farming community is non-competitive in the face of cheap imported foods. Moreover, food preferences have changed over the years from traditional cuisine to imported tastes, mimicking the popular foods in most developed countries. Behavior patterns have thus over-ridden local food "availability" in the food security equation. This exposes the islanders to high risk from external shocks, such as the food and fuel price crisis in 2008, with little realistic provision of local crop insurance owing to the high risk of extreme weather.

Moreover, Jamaica is experiencing serious diseases of coffee, citrus and banana, and loss of preferential trade status of the latter due to a WTO ruling. Even 80% of the goat meat is imported. Jamaica is nevertheless on course to become hunger-free, with its government trying to stimulate farmers to invest in their business, and the help of international partners (see Section 4.5.5).

In Samoa, the price of imported rice tripled after the 2008 food price crisis. More recently, a trend has been observed for islanders to start switching back to the varied traditional food, with the government encouraging this. The fruit and vegetable sector is characterized by subsistence farmers growing for home consumption on customary land, with very few commercial producers. There is scope for import substitution and exports for both livestock and crop produce. The government has a strategy in place, and is endeavoring to promote this through the Samoan Agricultural Competitiveness Enhancement Program (SACEP).

### 2.5.2 Individuals

Superimposed on *communities* most at risk of food and nutrition insecurity, from urban slums or refugees for example as discussed earlier, there is a subordinate risk of individual deprivation. Several categories of people comprise the most vulnerable of the marginalized in developing countries, subgroups of the rural and urban poor, who are not sufficiently, or at all, covered by national safety nets.

These individuals are often *children*, such as the millions of street children who have escaped from an abusive home environment or been thrown out of the home, had their mother/parents/carer die of disease or violence, or as child soldiers have been demobbed. It is estimated that there are one million street children just in Senegal, for instance, and another half million in Zambia. Many family units with houses in which to live are headed by a child, in the absence of parents.

Another high-risk group is *women* in general, especially widows or those who are pregnant and/or head of households following the desertion or illness/death of their partners. Each of these women may have difficulty in coping with the number of children in her charge, some of whom may not be hers, but whom she tries to care for, perhaps her grandchild whose parents have died or the mother working in a distant town. With such responsibilities she has little time to seek paid employment herself, or cultivate any land she may have. A woman of child-bearing age may not be able to prevent pregnancy, because of patriarchal social customs and/or she has neither the knowledge nor money to buy or use family planning methods, or has no access to them. Women at risk are numbered in hundreds of millions, such as those in Columbian refugee camps in Ecuador (Case Study 4 on the book's companion website), or in Kibera slum, Nairobi (Case Study 10).

The "grandmother syndrome", in which a young child is left in the care of the grandmother, or aunt in the impoverished rural village, is a common social cause of hunger and undernutrition of children. The grandmother (who is herself highly vulnerable to food and nutrition insecurity) has neither the energy nor opportunity to work to buy the food, collect firewood or pay for the gas, paraffin or charcoal, in order to cook for the grandchild. So, the latter is fed on the cheapest food available, which will be low in protein and vitamins. Similarly, the *working mother* in town, without a partner and living in a slum, leaves her child during the day playing in the dirt among the flies. The child inevitably gets intestinal infections which reduce absorption of the foods ingested. The malaria fever which the child contracts may not be properly treated and the condition, which leads to loss of appetite, results in undernutrition, if the child is spared death from the malaria itself, resistance to which is impaired because of its poor nutritional status—a vicious cycle of poverty and undernutrition.

Another group of people highly vulnerable to food insecurity is the *physically/mentally handicapped*, who cannot find work or enough of it, or cannot handle it. All age groups can clearly be affected. Often such individuals are casualties of war.

Many of the most vulnerable people and groups face numerous and overlapping constraints on their ability to cope with setbacks. For example, those who are poor and also from a minority group, or are female and have disabilities, face multiple barriers which can negatively reinforce each other.

## REFERENCES

Bergeron, G., Castleman, T., 2012. Program responses to acute and chronic malnutrition: divergences and convergences. Adv. Nutr. 3, 242–249.

Bryce, J., Coitinho, D., Darnton-Hill, I., Pelletier, D., Pinstrup-Andersen, P., 2008. Maternal and child undernutrition: effective action at national level. Lancet. 371 (9611), 510–526.

B'Tselem, 2009. B'Tselem's investigation of fatalities in operation cast lead. 6 pp. <www.btselem.org/download/20090909_cast_lead>.

Chambers, R., 1988. Vulnerability, coping and policy. Inst. Dev. Stud. (IDS) Bull. 20 (2), 1–7.

Davies, S., 1993. Are coping strategies a cop out? IDS Bull. 24 (4), 60–72.

De Onis, M., Blossner, M., Borghi, E., 2012. Prevalence and trends of stunting among preschool children, 1990–2020. Public Health Nutr. 15 (1), 142–148.

FAO, 2009. Food security and nutrition security: what is the problem and what is the difference? Summary of the FSN Forum Discussion No. 34, April 1 to May 15, 2009. 5 pp.

FAO, IFAD and WFP, 2014. The State of Food Insecurity in the World 2014: Strengthening the Enabling Environment for Food Security and Nutrition. FAO, Rome.

Frankenberger, T., Coyle, P.E., 1993. Integrating household food security into farming systems research/extension. J. Farming Syst. Res. Ext. 4 (1), 35–65.

Goel, M., Mishra, R., Gaur, D., Das, A., 2006. Nutrition surveillance in 1–6 years old children in urban slums of a city in northern India. Internet J. Epidemiol. 5 (1).

Jayarajan, M., 2014. Tackling Malnutrition in Urban Slums. Centre for Health Market Innovations (managed by Results for Development Institute), Washington DC.

Kaufmann, S., 2008. The Nutrition Situation in Northern Lao PDR—Determinants of Malnutrition and Changes After Four Years of Intensive Interventions (Thesis). Institute of Nutritional Sciences, Justus Liebig University, Giessen, 245 pp. <http://geb.unigiessen.de/geb/volltexte/2009/6904/pdf/KaufmannSilvia-2009-03-03.pdf>.

Keats, S., and Wiggins, S. Future diets: implications for agriculture and food prices. Overseas Development Institute (ODI) Report, London. January 2014.

Maxwell, D.G., 1996. Measuring food insecurity: the frequency and severity of coping strategies. Food Policy. 21 (3), 291–303.

Maxwell, S., Smith, M., 1992. Household Food Security: A Conceptual Review. IDS, University of Sussex, Mimeo.

Micronutrient Initiative *et al.*, 2009. Investing in the future: a united call to action on vitamin and mineral deficiencies. Global Report. Micronutrient Initiative, Flour Fortification Initiative, USAID, World Bank, UNICEF, GAIN (Global Alliance for Improved Nutrition).

Psaki, S., et al., 2012. Household food access and child malnutrition: results from the eight-country MAL-ED study. Popul. Health Metr. 10 (24), published online December 13, 2012 http://dx.doi.org/10.1186/1478-7954-10-24.

Rockefeller Foundation, 2011. Searchlight South Asia Newsletter: Tracking Urban Poverty Trends in India, Bangladesh and Pakistan. January 20, 2011.

Sheeran, J., 2008. The challenge of hunger. Lancet. 371 (9608), 180–181.

Stevens, G.A., et al., 2012. National, regional, and global trends in adult overweight and obesity prevalences. Popul. Health Metr. 10, 22.

Tulchinsky, T.H., 2010. Micronutrient deficiency conditions: global health issues. Public Health Rev. 32, 243–255.

WHO, 1995. Physical status: the use and interpretation of anthropometry. WHO Expert Committee, Technical Report Series 854. WHO, Geneva.

WHO, 2000. World Health Report, 2000. WHO, Geneva.

WHO, 2014. Micronutrient deficiencies; iron deficiency anaemia. Retrieved from: <www.who.int/nutrition/topics/ida/en/> (accessed 24.04.14.).

WHO, FAO, 2006. Guidelines on Food Fortification and Micronutrients. 376 pp.

Zimmerman, M.B., 2009. Iodine deficiency. Endocr. Rev. 30, 376–408.

Chapter | Three

# Causes of Food Insecurity

## 3.1 MULTIDIMENSIONAL ETIOLOGY

Section 2.5 has scoped the predisposition of some communities and individuals toward food and nutrition insecurity, this discussion touching on "causes" of the conditions. More detail is given below on the technical and social causes underlying these correlations, the causes often interrelated.

The world faces a new conjunction of several factors with a cumulative effect bearing little relation to the causes of the major food crises at the end of the 1970s and in the early 1980s, particularly in the Sahel and the Horn of Africa (EC, 2009) *op. cit.* The new factors include declining world food stocks, population explosion in many developing countries especially in urban areas, food and fuel price fluctuations, expansion of biofuel production from erstwhile "food crops", and climate change.

## 3.2 POVERTY AND INSUFFICIENT AWARENESS

The OECD (Organization for Economic Cooperation and Development) defines poverty as a state which "encompasses different aspects of deprivation that relate to human capabilities including consumption and food security, health, education, rights, voice, security, dignity and decent work". Poverty is not merely "income poverty" or "consumption poverty", but poverty of equitable opportunity. In the view of the Organization of American States, poverty for such peoples implies and includes undernutrition, unemployment, illiteracy (especially among women), environmental risks and limited access to social and sanitation services, including health services in general (OEA, 2000).

Hunger, undernutrition and food insecurity prevent destitute populations globally from escaping poverty in its widest meaning, because they reduce their capacities to study, work and take care of themselves and their families. The development of the poorest countries will remain an unachievable goal for as long as hunger and undernutrition remain, because they place a direct strain on the economic and human potential of these countries, often from



**Food Security in the Developing World. DOI: http://dx.doi.org/10.1016/B978-0-12-801594-0.00003-8**

one generation to another. Development necessarily implies improving the situation in terms of food security, which is vital in order to combat poverty.

Nearly half of the world's population, more than 3 billion people, live in poverty, on less than $2.50 a day. More than 1.3 billion of these live in extreme poverty, on less than $1.25 a day. Of the 3 billion, *1 billion are children, of whom 22,000 die each day because of poverty*, according to UNICEF. Poverty is especially prevalent in sub-Saharan Africa and South Asia. Poor people spend a greater proportion of their disposable income on food, buy poorer quality food and eat less frequently. They are particularly vulnerable to economic shocks, and rising and fluctuating food prices (Case Study 9 on the book's companion website, http://booksite.elsevier.com/9780128015940). As mentioned in Section 1.1, food insecurity and poor health are bicausally related, yet poverty is also bicausally associated with this relationship. The reasons for all these conditions are similar, including poorly functioning markets and insufficiency of required awareness—of knowledge and skills. Gender, age and ethnicity are often overarching influences on both food insecurity and poverty within the family and community (see Chapter 3: Causes of Food Insecurity).

To cite one country, an estimated 64–80% of Zambians live in poverty, predominantly in rural areas (Central Statistical Office, 2006). Between 1995 and 2006, extreme poverty declined by only 7%, to 51% (Government of Zambia, 2008). Zambia's Gini coefficient is 0.53, indicating an unequal income distribution (World Bank, 2008). Farmers largely farm on traditional land, thus having no collateral to offer to access finance. Financing institutions are reluctant to give loans to farmers so they may invest in their farms, hardly surprising as 37% of agricultural loans have been nonperforming, compared with 13% across other sectors.[1] This has led to poverty and hunger for around 80% of farmers who have less than 1 ha, the asset profile of a quarter of Zambia's farmers. As a result of poor farm productivity, and resultant insufficient food or wherewithal to buy more in, chronic undernutrition leaves 45% of Zambian children stunted (Government of Zambia, 2007).

The most common direct causes of undernutrition in slums globally include faulty infant feeding practices and child care practices in general, poor awareness of children's food requirements and absence of a responsible adult care giver, and appalling environmental/hygiene and sanitation conditions leading to impaired absorption of nutrients due to intestinal infections and parasites. The physiological causes of undernourishment are well understood, whereas the complex socioeconomic reasons to explain how and why it occurs are often community-based. The chain of causality can be very long, going back for generations, and sustainable solutions and beneficial outcomes hard to actually bring about even with the benefit of a complete understanding of socioeconomic causality.

[1] Report in 2011 on Zambia's agricultural finance market commissioned by the Zambia National Farmers' Union.

The fight against poverty has assumed priority across the development spectrum (EC, 2009) *op. cit.* One nutrition-related success story, from Brazil, underlines the association between socioeconomic status and child stunting. In Brazil's north-east, stunting decreased from 34% in 1986 to just 6% in 2006 (Lima et al., 2010). Comprehensive statistical models show that the scourge of chronic undernutrition can be rapidly reduced if income among the poor rises and simultaneously there is increased access to schools, clean water, sanitation and basic healthcare (Monteiro et al., 2009).

The global proportion of people living on less than $1.25 per day fell from 43.1% in 1990 to 22.2% in 2008. However, the food, fuel and financial crises since 2008 worsened the situation of vulnerable populations and impacted poverty reduction programs in some countries. An excellent summary of the prospects of removing global poverty is provided by the quotation in Box 3.1, set against MDG 1a (the eradication of extreme poverty).

The hunger imperatives enshrined in MDG 1 (as mentioned by World Bank in Box 3.1) are discussed in Chapter 2 "Manifestations and Measurement of Food Insecurity" and Case Study 8 on the book's companion website.

### Box 3.1 Prospects for the Eradication of Poverty

*The world will not have eradicated extreme poverty in 2015, but it will have met the Millennium Development Goal target of halving world poverty. The proportion of people in developing countries (those classified as low and middle income in 1990) living on less than $1.25 a day fell from 43.1 percent in 1990 to 20.6 percent in 2010 and reached a new low in five of six developing country regions. Except in South Asia and sub-Saharan Africa, the target was met at the regional level by 2010.*

*Further progress is possible—and likely—before the 2015 target date. Developing economies are expected to maintain GDP growth of 5.3–5.5 percent over the next two years, with GDP per capita growth around 4.2 percent. Growth will be fastest in East Asia and Pacific and in South Asia, which still have more than half the world's poorest people. Growth will be slower in sub-Saharan Africa, the poorest region, but faster than in the preceding years, quickening the pace of poverty reduction. According to these forecasts, the proportion of people living in extreme poverty will fall to 16 percent by 2015.*

*Based on current trends, around 40 percent of developing countries have already achieved the first Millennium Development Goal, and only 17 percent are seriously off track, based on the methodology used in the 2013 Global Monitoring Report. However, in sub-Saharan Africa up to a third of countries are seriously off track—meaning that they would be unable at current rates of progress to halve extreme poverty rates by 2030. Progress is also sluggish among countries classified as fragile, and in conflict situations and small states.*

*Data gaps remain and hinder the monitoring of progress. About a fifth of developing countries have not conducted a survey since 1990, the minimum requirement for monitoring progress when using national accounts data to interpolate or extrapolate survey data. By number of countries, the gaps are greatest in East Asia and Pacific and especially among small states and in fragile and conflict situations.*

**Source: From World Bank (2014). World Development Indicators 2014. Washington, DC: World Bank. doi:10.1596/978-1-4648-0163-1.**

It is true that among "poor communities" there may be a certain lack of awareness concerning ways and means to reduce poverty. Yet, a note of caution is required about how to do so, requiring a need for awareness and sensitivity by those from outside of a given community seeking to effect a change within it. This is especially true of those from developed countries with substantially different "norms", to ensure that change is brought about on a demand-led basis, as much as possible.

For example, while there is a marked level of measurable poverty in the Amazonian region of South America (Case Study 5 on the book's companion website), the predominant perception among the indigenous peoples away from urban centers, is that they are not "poor" but have a different lifestyle, more in harmony with nature, even though in western eyes this may be seen as a synonym for poverty (UNEP, ATCO and CIUP, 2009). For a similar perspective from Africa, on August 5, 2015, the BBC News website posted a 4-minute video clip entitled "Harvesting happiness on a Kenyan farm" (www.bbc.co.uk/news/magazine-33728876). In that clip, women of Nyakiambi village (Nakuru County) explain the fulfilment which their simple life brings them, and that they never buy food as they grow all they need on their smallholdings.

## 3.3 ENVIRONMENTAL DEGRADATION AND CLIMATE CHANGE

Climate change is cyclic rather than a new phenomenon, though humanity is affecting climate now more than ever before, because of our numbers and lifestyle. Some 6,000 years ago, as the Sahara dried up, peoples migrated from there to the Nile Valley. Subsequently, some 4,300 years ago during the reign of Pharaoh Pepe II, ancient Egypt started to decline due to much reduced flow in the Nile, in turn due to less rain in the Ethiopian highlands, which was itself likely due to changes in oceanic tide patterns far away. Hieroglyphs on contemporary *stelae* show the catastrophe in the lower (northern) Nile. Adults ate their own children to survive, engulfed as they were by desert which could not sustain them. Irrigation was then only possible in Luxor and upstream of it, to which area residents of the lower Nile swarmed hoping to source food. The previously plentiful water transport system downstream had once enabled the pyramids to be built, and for a thousand miles of land along the Nile to be under irrigated farming.

Such environmental and cultural collapse happened not only in lower Egypt but also in the wider Western Asia region in which great civilizations had developed and flourished on the back of abundant irrigation water from other major river systems—the Akkadian empire in modern day Iraq and

Iran, Palestine, Turkey and the Greek mainland for instance, in which half a dozen highly advanced civilizations came to an abrupt end. When the Ethiopian rains returned, the Egyptian Middle Kingdom civilization which arose from the ashes of the old had a different governance structure, as the Pharaohs had been shown not to be all-powerful Gods after all, unable to avert the drought.

More recently, in the Ethiopian uplands, when the human population was less and rains more plentiful than they now are, the sustainable land management practice was slash-and-burn, one household having some 25 acres (10 ha) at its disposal. Now, with some 80 million people in Ethiopia, and a household's access to arable land being only around one acre (0.4 ha), the threat of land degradation is inevitable, through overexploitation and improper husbandry, and this degradation was in part responsible for the famines of the past. The ongoing restorative Productive Safety Net Program (PSNP) intervention in Ethiopia has proven exemplary in simultaneously restoring degraded land, enabling a greater carrying capacity of both humans and livestock, providing a social transfer safety net and building resilience to food insecurity.

That watershed management provides the core guiding principle for technical interventions of PSNP is admirable, in one of the world's most degraded environments, thereby fostering long-term sustainability of the environment. A new approach to land use has been long overdue. PSNP has brought this approach, through targeting communal land assets and underpinning enhanced and sustainable production levels thereon (see Case Study 1 on the book's companion website).

Severe environmental degradation has occurred in many other countries too, such as Somalia, due to high pastoral stocking rates in the context of a drying climate, together with the cutting of trees for firewood and charcoal. Denudation of the sandy soils has led to water and soil erosion when the rains do fall, great gullies having been formed. The largely barren surface is prey to wind erosion too. Because the rain water is not retained by the vegetation and planes down the slopes, the water table has fallen, making it more difficult for replacement trees to establish.

A report from several US universities in 2015 points to a Doomsday scenario, as a result of environmental degradation which humankind is inflicting on our planet (Ceballos et al., 2015). One example given is that lemurs in Malagasy forests are under threat of extinction, as a result of illegal logging and hunting for bushmeat. The importance of bushmeat as an erstwhile sustainable source of protein in the diet for many societies is increasingly under threat (see also Case Study 5 on Amazonia on the book's companion website).

A more detailed consideration of modern climate change and its influence on food security is given in Section 6.9.

## 3.4 FOOD PRICE HIKES AND PRICE INSTABILITY

The food price hikes and their instability since 2008 have been largely attributed to drought, world population growth (see Section 6.4), increased demand by the growing middle class in China and India, rise in international fuel prices, depreciation of the US dollar and global economic crises, and trade-related factors such as export restrictions and hoarding/market manipulation. The food price crisis of 2008 and beyond is discussed in detail in Case Study 9 on the book's companion website.

## 3.5 CONFLICTS

On June 18, 2015, UNHCR issued its chilling annual report for the previous year, *Global Trends 2014: World at War*, in which it says that 59.5 million people are currently forcibly displaced by war and persecution, compared with 51.2 million a year previously. Half of these are children. Since early 2011, the main reason for the acceleration has been the war in Syria, now the world's largest driver of displacement. Over the previous 5 years, at least 15 conflicts erupted or reignited—8 in Africa, 3 in the "Middle East", and 1 in Europe. One in every 122 human beings is now a refugee, an IDP or seeking asylum. "We are witnessing a paradigm change, an unchecked slide into an era in which the scale of global forced displacement as well as the response required is now clearly dwarfing anything seen before", the UN High Commissioner for Refugees António Guterres said. "It is terrifying that on the one hand there is more and more impunity for those starting conflicts, and on the other there is seeming utter inability of the international community to work together to stop wars, and build and preserve peace".

An example of long-term food insecurity arising from the two Liberian civil wars has been given in Chapter 1 "Introduction". Other examples of conflict which could be mentioned are in Eastern Africa—South Sudan and Eritrea. South Sudan is undergoing ethnically based conflict. Warnings were issued mid-May 2014 by FAO, humanitarian relief organizations and the Famine Early Warning Systems Network (FEWS NET), about the impending threat of famine "within 3 or 4 months" if the government and rebel forces did not stop fighting. The provisional results of Integrated Food Security Phase Classification (IPC) analysis suggested that food security had dropped to alarming levels in the three conflict-affected states of Jonglei, Unity and Upper Nile. The deteriorating food insecurity situation hits hardest during the period of the area's preharvest "lean season" when "normal" food insecurity is at its peak.

As for the second example, a 2-year war broke out between Eritrea and Ethiopia in 1998, started by the Eritrean government in a miscalculated military adventure, in which tens of thousands died. Gash Barka, the land border between the two countries, is a semi-arid area. The precarious agricultural livelihood that existed there before that war was totally disrupted by the cross-border conflict, and the area became contaminated by landmines.

That straight "poverty" degraded into *structural* poverty was, in part, because camels, which were the source of animal draft for plowing farmers' fields were looted, eaten, blown up by landmines or fled the conflict zone. Even when rain does come to Gash Barka, because of the near absence of camels, the land cannot be tilled and planted sufficiently to feed the communities. Serious food insecurity has resulted, warranting large-scale humanitarian aid. Poverty in the region is so great, that many people there cannot afford to buy replacement batteries for their radios, which would enable them to listen to the news or government announcements on how they may relieve their poverty. Unfortunately, Eritrea is close to the bottom of the UNDP Human Development Index, and the repressive regime has caused thousands to flee, all so un-necessary following the successful independence war against Ethiopia which culminated in 1991.

## 3.6 WEAK INSTITUTIONAL ENVIRONMENT

### 3.6.1 Uncoordinated Policies and Institutions

In regard to uncoordinated policies and institutions, both international and national perspectives apply. Regarding *international* institutions, progress has been made over past years in West Africa for instance, thanks to the work of existing platforms—such as the "*Réseau de Prévention des Crises Alimentaires*" (RPCA) and the "*Comité permanent Inter-Etats de Lutte contre la Sécheresse au Sahel*" (CILSS). Building upon lessons learned, approaches have been developed to better face food crises. The improved functioning of early warning systems has helped countries and donors better anticipate the ongoing crisis.

The "*Alliance Globale pour l'Initiative Résilience*" (AGIR) Partnership, launched by the European Commission on June 18, 2012, provides a resilience roadmap, building on and reinforcing existing regional strategies—such as the joint regional strategy of the Economic Community of West African States (ECOWAS), the *Union Economique et Monétaire Ouest Africaine* (UEMOA) and the CILSS, with the support of the Sahel and West Africa Club (SWAC). A regional Plan of Action to strengthen resilience in a permanent and sustainable way in the Sahel, drawn up by the West African regional organizations with the support of the donor community, was presented to a high-level meeting of all the States concerned in Ouagadougou, Burkina Faso, on December 6, 2012. In a Joint Statement outlining the principles, priorities, and next steps of the Alliance, stakeholders pledged to achieve the objective "Zero Hunger" within the next 20 years (see Section 2.1.1). The Alliance aims at building resilience of vulnerable populations by consolidating responses to food crises and structural causes of food insecurity, as well as to chronic undernutrition. In line with the Paris Declaration on Aid Effectiveness, AGIR is founded on the principles of West African leadership, and harmonization and coordination amongst partners.

Regarding *national*-level policy and institutional upgrading, an example related to nutrition is given here. Nutrition-related policies, laws and institutions are crucial for achieving progress on nutrition security. Though having policies and laws in place does not guarantee implementation, they nevertheless indicate a government's public commitment and offer an entry point for civil society engagement. The 2014 Global Nutrition Report (IFPRI, 2014), by an independent expert group guided by a stakeholder group, *inter alia* offers guidance on how to build such commitment and accountability. The report aims to empower "nutrition champions" at the national level to better inform policy decisions and strengthen the case for increased resources.

One of the premises of the Scaling Up Nutrition (SUN) movement, which has 54 member countries, is that efforts to reduce undernutrition need to become more coordinated across sectors and stakeholders, and more aligned with results frameworks (see Section 4.7.3). Accordingly, SUN members are pioneering a new way of assessing institutional transformation for nutrition improvement. Thirty-seven countries have conducted self-assessments of their annual progress in relation to four processes defined in the 2012–15 SUN Movement Strategy, these being: bringing people into a shared space for action; ensuring a coherent policy and legal framework; aligning actions around a common results framework; and, tracking finances and mobilizing resources. The self-assessment scores show that ongoing efforts to coordinate multiple stakeholders, develop policies and legislation, and mobilize resources for nutrition have yet to be fully translated into properly managed and monitored actions, and into investments that are scaled up, aligned and adequately accounted for.

The 2014 Global Nutrition report records that 77 countries have data for five of the six SUN undernutrition policy and legislation indicators, though Brazil is the only country with a top score in all five. The report notes that China and Thailand, both recognized for their strong performance in reducing undernutrition over the past 20 years, do not have a top score in any of the indicators.

### 3.6.2 Insufficient Investment in Agriculture and Public Services

In general, there has been inadequate investment in agriculture in developing countries. Funding smallholder agriculture can be an unrewarding initiative unless it is done in a comprehensive way. Bringing high-yielding improved varieties to the farmers will unlikely bring sustained rewards, unless supplemented with complementary actions, as was done during the early Green Revolution (see Section 5.3.1.2). Commercial high input market-led agricultural ventures also demonstrate that not only can plentiful good-quality food be produced, but that this can be done profitably and improved living standards thereby assured.

Similarly, investment in public services like roads and drainage, water supply and sanitation, health services and housing can all contribute to the simultaneous removal of squalor, urban and rural decay, food and nutrition insecurity. There is a very strong positive correlation between water security and nutrition security.

Owing to modest available budgets of many governments in the developing world, Public-Private-Partnerships (PPPs) offer a win-win collaborative opportunity to improve development outcomes, eg, improved food availability and economic access. PPPs can enable the public sector to harness the expertise, efficiencies and financing of the private sector (with the latter gaining access to a market for its services and means for cost recovery and profit), and benefit through tariffs levied on consumables or services sold. An example from the Caribbean is the Puerto Rico Expressways (*Autopistas de Puerto Rico S.E*) and *Societas Europaea*, which together operate the Teodoro Moscoso Bridge on behalf of the Puerto Rico Highways and Transportation Authority. The toll bridge spans the San José Lagoon linking the municipalities of San Juan and Carolina, and provides better connectivity to the Luis Muñoz Marín international airport on both sides of the lagoon. In neighboring Central America, the potential of PPPs for plantation worker village upgrades in Belize is discussed in Case Study 3 on the book's companion website.

A successful example from South Africa is the PPP to upgrade and renovate Pelonomi and Universitas hospitals' service delivery in Bloemfontein, Free State, to address MDG 6, the combat of HIV/AIDS, malaria and other diseases, thereby indirectly improving food and nutrition security. Funding for this venture by the Provincial government alone was insufficient. Two healthcare companies work together as the private sector partner.

## 3.7 PREDISPOSITION OF THE COMMUNITY TO DISEASE AND INTESTINAL AFFLICTION

Onset of food and nutrition insufficiency in an individual is better combatted if that person is of sound constitution. If the body reserve "buffer" is already depleted and diminished from the "norm" (through diseases like malaria, cholera or HIV/AIDS or intestinal parasites), it takes less time to reach a morbid physical condition due to insufficient nutritious food intake or retention. Through loss of appetite and compromised ability to work the fields or earn off-farm, the infection or infestation also predisposes the individual to a more rapid onset of undernutrition.

Living conditions which facilitate such disease, infection and infestation are associated with poverty, as described in Chapter 2 "Manifestations and Measurement of Food Insecurity" and Section 3.2, and Case Studies 3 and 10 on the book's companion website.

## 3.8 LARGE-SCALE LAND LEASE BY ONE COUNTRY IN ANOTHER

Some countries which have insufficient land or water resources to produce the food they need have recently started to lease tracts of land in other countries, which are not constrained by shortages of land or water. The purpose is to make good their shortfalls through securing a source, one which is immune to commodity price fluctuation on world markets. Corporate investors are also engaging in such leasing, with a profit motive. Since 2009, reports surfaced of such large-scale long-leasing of agricultural land by investors or offshore governments (especially from Asia). The scale of this land leasing is staggering. Over the period 2008–09 alone, 56 million hectares of land worldwide was contracted in this way, compared with only 4 million hectares prior to 2008. Two-thirds of this land has been in Africa, in countries like Liberia, Ethiopia, Sudan, Mozambique, Uganda and Tanzania. Whether such arrangements are really to the benefit of the host countries and farmers who were farming the land previously has become the subject of intense debate[2] (World Bank, 2011; Schaffnit-Chatterjee, 2012).

Some of the *positive* perspectives include: incremental paid employment and income generated within the host communities; urban migration stemmed; concessionary fees charged by host governments; improved technologies to enhance productivity imported and applied, and value added from agriprocessing and better marketing.

By contrast, there are potential *negative* consequences of such large-scale land leasing, as has long been known since the classic study on oil palm concessions in Liberia (Clower et al., 1966), in which there is national growth *without* development. These negatives have led to the term "land grabbing" being used, and can comprise a threat to food and nutrition security, for reasons including: land and water resources of poor villagers, traditionally used to support smallholder livelihoods, forcibly expropriated without due process or consultation by heavy-handed governments, local rights being ignored; environmentally diverse smallholdings replaced by monocropping regimes, with attendant agrochemical pollution and loss of biodiversity; the incremental jobs provided are few, temporary and poorly paid; due to poor business planning or local institutional unpreparedness, some large projects are non-viable; tax revenues to government are low due to huge tax concessions offered to investors; dubious contribution to poverty reduction and sustainable development in the host country.

These negatives need not arise, if local land rights are better-secured, and institutional and regulatory capacity improved through investment better targeting smallholder agriculture and cooperatives (rather than vast plantations), so that both partner and host countries can simultaneously improve their food security. *Investment is better made in farmers rather than farmland, involving a participatory approach by investors and governments, and the rights of local landholders enshrined in the terms of the lease following proper consultation and due process* (also see Section 6.8 on food sovereignty).

---

[2] Article by Lorenzo Cotula of the International Institute for Environment and Development (IIED) on the BBC website of February 22, 2012 (www.bbc.co.uk/news/world-africa-17099348).

## 3.9 LARGE AREAS OF ARABLE LAND SET ASIDE FOR BIOFUEL PRODUCTION

### 3.9.1 Introduction

A significant proportion of harvest from land devoted to some major traditional food crops is not used for food, but processed into "alternative energy" fuels for vehicles—the fermentation of mainly corn (maize) for ethanol, sugarcane for petrol/gasoline and soya or other vegetable oil and animal fats processed for diesel. In 2010, the USA and Brazil between them produced around 90% of the world's bioethanol, while the EU produced some 53% of the world's biodiesel. In the United States, 40% of the corn—some 15% of global corn supplies—is now diverted from its traditional use for human food and feedstuff for livestock, to make ethanol, up from just 5% in the year 2000.

Two contrasting narratives have arisen as to the wisdom of biofuels as they affect food security in the developing world, with powerful lobby groups on both sides of the debate, and a passionate debate it is (the position of the World Bank is ambiguous, depending on who voices an opinion). A short survey of that debate will be presented below, though without a definitive conclusion being possible on which view, if either, is likely to be proven the more convincing by posterity.

### 3.9.2 The Anti-Biofuels Lobby

Some analyses have identified negative impacts of biofuels on poor farmers and their communities, either directly in the form of land expropriations or indirectly through the concentration of resources on large-scale farming operations. It has been shown by Ortiz et al. (2013) that biofuel crop replacement of forests can seriously impact ecosystem services provision and food security for rural communities, through a reduction in livelihood opportunities or change in land tenure patterns, for example (see Case Study 5 on the book's companion website). The recent food price spikes (2008, 2011–12), together with a wide range of studies, also implicate biofuels as a key driver of price volatility.

Mitchell (2008) of the World Bank reported that biofuel production played a significant role in food price hikes. Mitchell examined the factors behind the rapid increase in internationally traded food prices since 2002 and estimated the contribution of various factors such as increased production of biofuels from food grains and oilseeds, the weak dollar and the increase in food production costs due to higher energy prices. He concluded that large increases in biofuels production in the United States and Europe were the main reason behind the steep rise in global food prices, though Brazil's sugar-based ethanol did not push food prices appreciably higher.

The World Bank's Agricultural Unit senior economist, Robert Townsend, explaining the major reasons for recent global food price increase, noted that surging demand for food crops increased faster than supply due primarily to biofuel policies in industrialized countries and to a lesser extent changing diets in rapidly growing developing countries. Furthermore, he opined that those policies have diverted food crops from traditional export markets to production of ethanol and biodiesel.[3]

Wright (2012) also argued that low grain stocks during 2007/08 at a time of strong biofuel demand and increased incomes by China and India were the key causes of the post-2007 grain price increases. A paper by Zilberman et al. (2012) showed that although not the dominant contributor, biofuels *had* caused a rise in food commodity prices, the size of the effect depending on the type of biofuel.

A report by HLPE (2013) affirmed the negative impacts of biofuels to date and recommended decisive action. The report pointed out that in less than one decade, world biofuel production has increased five times, from less than 20 billion L/year in 2001 to over 100 billion L/year in 2011. The steepest rise in biofuel production occurred in 2007/08, concomitantly with a sharp rise in food commodity prices (HLPE, 2011) *ibid.* and accompanied by food riots in cities of many developing countries. In comparison with average food prices between 2002 and 2004, globally traded prices of cereals, oils and fats have been on average from 2 to 2.5 times higher in 2008 and 2011–12, and sugar prices have had annual averages of from 80% to 340% above their 2000–04 levels. These price increases were accompanied by price volatility and price spikes to an extent unprecedented since the 1970s. Though other factors have been adduced in the many studies that have since been dedicated to the issue of rising food prices (HLPE, 2011), the steeply rising demand for the production of biofuels was identified as an important causative factor by many observers and a wide range of organizations, from civil society organizations (CSOs) to the World Bank (2008).

The HLPE paper of 2013 points out that many factors influence world supply and demand for food, concomitantly with biofuels. What mattered most for the authors of that report and analysis was not the net overall effect of all factors on the net food price, which had been addressed, for example, in HLPE (2011) *ibid.*, but the isolated effect of biofuels on food prices, *everything else being equal.* A central challenge was to disentangle and separate the impact of biofuels from all the other factors so that it could be analyzed from the standpoint of its *additional* impact, which leads to *additional* price effects.

A contemporary report from ActionAid USA (2013) *op. cit.*, prepared in collaboration with Tufts University, citing extensive economic modeling research on biofuels and several other challenges to global food security, confirms that one of the main threats to the world's ability to feed itself in future is the continued expansion of first generation biofuels. This report claims that biofuels expansion is a relatively recent phenomenon that has been poorly captured by most economic modeling to date. With national mandates and targets significantly driving biofuels expansion, updated forecasts are urgently needed to help policymakers assess the food security implications of

[3] Retrieved from http://go.worldbank.org/KIFNA6OZS0 (accessed Oct. 22, 2013).

current policies. Existing policies result in rising and more volatile food prices, with up to 40% of recent price increases in agricultural commodities attributable to biofuels expansion, in ActionAid's view. Moreover, the report believes that those policies are projected to divert as much 13% of cereal production from needed food production by 2030. ActionAid USA opines that the world is not waiting for the Committee on World Food Security (CFS) to lead in this debate! (NB. CFS was set up in 1974 as an intergovernmental body to serve as a forum for review and follow up of food security policies—see Section 3.9.4).

Though it may be that markets rather than policies are the stronger driver of biofuel production levels, policymakers around the world are beginning to contend with food-fuel competition. The HLPE (2013) *ibid.* report states that by mid-2013, more than 50 countries have adopted biofuels policies in which food security has become a central conditioning criterion for biofuels promotion, with explicit policies in China, India and South Africa not to base biofuels on food crops or on land used for food. The US Congress is under pressure to reform, or even repeal, its biofuels mandate, and the EU recently reduced its own mandate by half, explicitly recognizing the negative impacts of food-based fuels.

### 3.9.3 The Pro-Biofuels Lobby

Set against the negative narrative, proponents regard biofuels as a strategy to mitigate climate change, to provide new opportunities for income and employment generation, for both primary crop producers and processors, and to bring in much-needed capital, technology and knowledge to developing country agriculture. Biofuel crops are being promoted in the Peruvian Amazon as an alternative to coca plantations, where oil palm plantations are replacing forests (Khwaja, 2010), and provide a short-term carbon payback in highly degraded Amazonian rangelands (Gibbs et al., 2008).

Despite the case made above for biofuel production driving grain price hikes, a study by Baffes and Haniotis (2010), from the same World Bank Policy Unit mentioned previously which produced the Unit's previous study of 2008, concluded that that earlier paper may have overestimated the contribution of biofuel production as the effect of biofuels on food prices had not been as large as originally thought, and that the use of commodities by financial investors may have been partly responsible for the 2007/08 spike.

A more recent paper still by the World Bank (Baffes and Dennis, 2013), also backtracked from the original 2008 position (this reported on gleefully by the trade website www.biofuelsdigest.com). These authors concluded that most of the price increases can be accounted for by crude oil prices (more than 50%), followed by stock-to-use ratios and exchange rate movements, which are estimated at about 15% each. Baffes and Dennis opined that worldwide, biofuels account for only about 1.5% of the area under grains/oilseeds, this raising serious doubt about claims that biofuels account for a big shift in global demand. Even though widespread perceptions about such a shift played a big role during the recent commodity price boom, they noted that corn prices hardly moved during the first period of increase in US ethanol production, that oilseed prices dropped when the EU increased its use of biodiesel, and that prices spiked when ethanol use was slowing in the United States and biodiesel use was stabilizing in the EU.

### 3.9.4 Conclusions of the Debate

So what is to be made of these conflicting positions ? The UN CFS is mandated to be the principal international agency coordinating global responses to the food price crisis, and dealing with the realities of the integration of food markets with fuel and financial markets. In October 2011, the CFS had recommended a "*review of biofuels policies—where applicable and if necessary—according to balanced science-based assessments of the opportunities and challenges that they may represent for food security so that biofuels can be produced where it is socially, economically and environmentally feasible to do so*". In line with this, the CFS requested the HLPE to prepare an Expert Report (HLPE, 2013) *ibid.*, to "*conduct a science-based comparative literature analysis taking into consideration the work produced by the FAO and Global Bioenergy Partnership (GBEP) of the positive and negative effects of biofuels on food security*".

The findings of this Expert Report have been mentioned earlier. Because of its concern at the threat posed by biofuels to global food security, the CFS put the issue of biofuels and food security on the agenda of the 40th CFS meeting in Rome in October 2013, decisions taken there to be informed by the Expert Report published earlier that year. At that CFS meeting, more than 80 organizations from around the world presented an open letter urging the Committee to take action against such a significant amount of arable land being devoted to biofuel crops.

As reported by the news website of Al Jazeera (2013), members of CSOs present at that meeting said in a press release "Small scale food producers have spoken powerfully here about the reality they are confronted with every day: that biofuels crops compete with their (smallholder's) food production, for the land they till and for the water that sustains them", and accused the CFS of defending the interests of the biofuels industry and legitimizing violations of the right to food. The news website opined that the proposal tabled that biofuels policies which harm food security should be reformed, was rejected by the CFS. So too, was any mention of the land and water impacts of runaway biofuels expansion. There was also no acknowledgment of the negative impacts of biofuel policies and mandates in the United States and European Union, which have been instrumental in stimulating and sustaining the biofuel industry.

CSOs at the CFS meeting saw conflict of interests at play, with lobby groups representing powerful industry interests, from the biofuels companies themselves and agribusiness firms capturing the benefits of high prices and subsidized demand for their products exerting pressure from behind the scenes. The most powerful countries at the negotiating table were the same ones benefitting from the "burning of food in our cars" (Al Jazeera, 2013) *ibid.* Canada and the United States in particular "played the loudest", supported by the EU, Brazil and Argentina. Only South Africa, a lonely voice, joined with civil society to speak up for the "victims of these policies".

Those who lament the rise of the biofuel industry point to multiple and severe food security outcomes, as competition for land and water has a direct impact on the cost of food. In 2008, global food prices doubled, hurting poor consumers in particular around the world. Most dramatically, biofuel producers are seen to have been key drivers of domestic and large-scale land investments/"land-grabbing" acquisitions in African and other developing countries.

Recommendations of the HLPE (2013) paper include:

- Food security policies and biofuel policies cannot be separated because they mutually interact. Food security and the right to food should be priority concerns in the design of any biofuel policy.
- Governments should adopt the principle: biofuels shall not compromise food security and therefore should be managed so that food access or the resources necessary for the production of food, principally land, biodiversity, water and labor are not put at risk. The CFS should undertake action to ensure that this principle is operable in the very varied contexts in which all countries find themselves.
- Given the trend toward the emergence of a global biofuels market, and a context moving from policy-driven to market-driven biofuels, there is an urgent need for close and proactive coordination of food security, biofuel/bioenergy policies and energy policies, at national and international levels, as well as rapid response mechanisms in case of crisis.
- There is also an urgent need to create an enabling, responsible climate for food and non-food investments compatible with food security.
- Governments must adjust biofuel policies and devise mechanisms to prevent (market-driven) biofuel demands posing a threat to food security from price rises and diminishing access to land and associated resources for food.
- Governments should ensure that the principles for responsible investment in agriculture, currently being elaborated by the CFS, will be effectively implemented and monitored, especially in the case of investments for biofuel production.
- The principles of free, prior and informed consent and full participation of all concerned in land-use investment, should be used as preconditions for any land investments.

A paper by Bailey (2012) pointed out that in the United States the then-ongoing drought had refocused attention on ethanol, production of which then used some 40% of US corn production. Ethanol's demand for corn does not adjust to supply shocks such as this. It is set by government mandate so is perfectly inelastic, forcing adjustment instead on the livestock sector and food consumers, amplifying price spikes in the process. Calls have therefore mounted for the mandate to be relaxed so that ethanol bears its fair share of the adjustment burden. The benefits of doing so could be significant. Modeling by the UK government had found that halving the US mandate during a price run-up could reduce the magnitude of the spike by about 40%.

However, Bailey observed that the US administration has rejected this option. 2012 was an election year, and maybe the ethanol mandate was too valuable a subsidy to the politically crucial corn-growing states to be threatened. But the question is likely to return soon, as further price spikes seem inevitable. Global food stocks remain close to crisis thresholds and are struggling to recover as demand continues to outstrip production growth, markets remain tightly balanced and no agreements exist to prevent or limit export bans. *In these circumstances the world is only one or two bad harvests away from another global crisis, with biofuel mandates increasingly hard to justify.*

## REFERENCES

ActionAid USA, 2013. Rising to the challenge: changing course to feed the world in 2050 (Wise T.A., Sundell K.), 28 pp. <http://www.actionaidusa.org/publications/feeding-world-2050>.

Al Jazeera, 2013. Fiddling in Rome while our food burns. Article on its website of Oct. 17, 2013. Retrieved from: <http://www.aljazeera.com/indepth/opinion/2013/10/20131015132924285478.html> (accessed 17.10.13.).

Baffes, J., Dennis, A., 2013. Long-term drivers of food prices. Policy Research Working Paper 6455. World Bank Development Prospects Group, and Poverty Reduction & Economic Management Network, 37 pp.

Baffes, J., Haniotis, T., 2010. Placing the 2006/08 commodity price boom into perspective. Policy Research Working Paper 5371. Development Prospects Group, World Bank, 42 pp.

Bailey, R., 2012. Are We Facing Another Global Food Price Crisis? Chatham House Royal Institute of International Affairs, London.

Ceballos, G., Ehrlich, P.R., Barnosky, A.D., García, A., Pringle, R.M., Palmer, T.M., 2015. Accelerated modern human-induced species losses: entering the sixth mass extinction. Sci. Adv. 1, e140025.

Central Statistical Office, 2006. Living Conditions Monitoring Survey 2006. Lusaka, Zambia.

Clower, R., Dalton, G., Harwitz, M., Walters, A.A., 1966. Growth Without Development—An Economic Survey of Liberia. Northwestern University Press, Evanston, IL, 385 pp.

EC, 2009. Food Security: Understanding the Meaning and Meeting the Challenge of Poverty. EuropeanAid Cooperation Office, Brussels, 28 pp., p. 8.

Gibbs, H.K., Johnston, M., Foley, J.A., Holloway, T., Monfreda, C., Ramankutty, N., et al., 2008. Carbon payback times for crop-based biofuel expansion in the tropics: the effects of changing yield and technology. Environ. Res. Lett. 3, 034001. Available from: http://dx.doi.org/10.1088/1748-9326/3/3/034001.

Government of Zambia, 2007. Zambia Demographic and Health Survey. Lusaka.

Government of Zambia, 2008. Millennium Development Goals Progress Report. Lusaka.

HLPE, 2011. Price volatility and food security. A Report by the High Level Panel of Experts on Food Security and Nutrition of the Committee on World Food Security, Rome.

HLPE, 2013. Biofuels and food security. A Report by the High Level Panel of Experts on Food Security and Nutrition of the Committee on World Food Security, Rome 2013. 132 pp.

IFPRI, 2014. Global Nutrition Report 2014: Actions And Accountability to Accelerate the World's Progress on Nutrition, Washington, DC. 117 pp.

Khwaja, A., 2010. Explaining the context of the Bioenergy and Food Security Project (BEFS) in Peru. Environment and Natural Resources Working Paper 40. FAO, Rome.

Lima, A.L., Silva, A.C., Konno, S.C., Conde, W.L., Benicio, M.H., Monteiro, C.A., 2010. Causes of the accelerated decline in child under-nutrition in Northeastern Brazil (1986–1996–2006). Rev. Saude Publica. 44, 17–27.

Mitchell, D., 2008. A note on rising food prices. Policy Research Working Paper 4682. World Bank, Washington, DC.

Monteiro, C.A., D'Aquino Benicio, M.H., Conde, W.L., Konno, S., Lovadino, A.L., Barros, A.J.D., et al., 2009. Causes for the decline in child under-nutrition in Brazil, 1996–2007. Rev. Saude Publica. 43, 35–43.

OEA, 2000. Segundo informe sobre la situación de los derechos humanos en el Peru. Comisión Interamericana de Derechos Humanos, Document 59. Organização dos Estados Americanos.

Ortiz, R., Nowak, A., Lavado, A., Parker, L., 2013. Food Security in Amazonia: Report for Global Canopy Program and International Center for Tropical Agriculture as part of the Amazonia Security Agenda Project, 89 pp.

Schaffnit-Chatterjee, C., 2012. Foreign investment in farmland: no low-hanging fruit. Current Issues—Natural Resources. Deutsche Bank Research, 23 pp.

UNEP, ATCO and CIUP, 2009. Environment outlook in amazonia 'Geo Amazonia' (page 75). UN Environment Programme, Panama. Amazon Treaty Cooperation Organization, Brazil (www.unep.org/pdf/GEOAMAZONIA.pdf), 166 pp.

World Bank, 2008. Poverty Report 2007, Washington DC, USA.

World Bank, 2011. Rising global interest in farmland: can it yield sustainable and equitable benefits? (Deininger, K., Byerlee, D. et al.) Agricultural and Rural Development Publication 59463, 266 pp.

World Bank, 2014. World Development Indicators 2014. World Bank, Washington, DC. Available from: http://dx.doi.org/10.1596/978-1-4648-0163-1, License: Creative Commons Attribution CC BY 3.0 IGO. 136pp. page 2.

Wright, B.D., 2012. International grain reserves and other instruments to address volatility in grain markets. World Bank Res. Obser. 27, 222–260.

Zilberman, D., Hochman, G., Rajagopal, D., Sexton, S., Timilsina, G., 2012. The impact of biofuels on commodity food prices: assessment of findings. Am. J. Agr. Econ. 95 (2), 275–281.

Chapter | Four

# Mitigation of Current Food Insecurity

## 4.1 COPING STRATEGIES AT MICROLEVEL

The range of coping strategies at microlevel, whereby vulnerable households and individuals endeavor to remain food secure is considered here. Davies (1993) *op. cit.* makes the distinction between "coping strategies" (fall-back mechanisms to deal with a short-term insufficiency of food) and "adaptive strategies" (long-term or permanent changes in the way in which households and individuals acquire sufficient food or income). Constructive "adaptive strategies" for cropping are mentioned in Section 5.3.1.

Examples of coping strategies include reducing or rationing consumption (relatively minor changes such as eating a less expensive/less preferred food, or severe changes such as going for an entire day without eating); altering household composition; altering intrahousehold distribution of food; depletion of grain stores; increased use of credit for consumption purposes; increased reliance on food hunted or gathered from the wild; short-term alterations in crop and livestock production patterns; mortgaging and sales of assets; short-term able-bodied labor migration or distress migration of whole families. A distinction also needs to be made between "coping strategies" and "strategies which fail to cope", the latter sometimes referred to as "negative coping strategies" (Drèze and Sen, 1989)—such as, in order to buy food, selling the thatch from the family house to a neighboring family as fodder for its starving livestock.

Under the Ethiopia PSNP, evidence from the 2008 impact evaluation and other sources suggests an overall drop in negative coping strategies among PSNP participants who received *predictable high-value* transfers, compared with non-PSNP participants (see Case Study 1 on the book's companion website, http://booksite.elsevier.com/9780128015940). Building resilience (see Section 5.1.3) is a key goal in development work, such that negative coping strategies are not needed.

The number of chronically undernourished people jumped from an already grave 854 million in 2007 to a record 1.02 billion in 2009, with the



Food Security in the Developing World. DOI: http://dx.doi.org/10.1016/B978-0-12-801594-0.00004-X

poorest, landless and female-headed families hit the hardest. Poor people—who typically spend 50–70% of their household income on food—were hit badly by the FPC, as recorded by IFPRI (2008). In the cases studied, across Asia, Africa and Latin America, families cut back on food, sold assets and sacrificed healthcare and education. Women and girls increasingly ate last and least, some sold sex for food and the elderly were abandoned. Maxwell (1996) *op. cit.* considers how the coping strategies employed might serve as indicators to measure the level of food insecurity.

*The human face of global food price rises is often missing among the abstract discussions of macro-economic trends and global food price indices.* In order to understand the impact of the rise in global food prices through much of 2010 and into early 2011, Oxfam and partners from the Institute of Development Studies spoke with people affected in Bangladesh, Indonesia, Kenya and Zambia. Their research identified the coping strategies of people forced to eat less and shift to cheaper, less preferred and often poorer-quality foods, and less diverse diets. They also sold productive assets, incurred debt, withdrew children from school, married early, worked harder, lived more frugally, managed on a day-to-day basis and *in extremis* migrated to areas where food was thought to be more available (Hossain and Green, 2011).

The effects are gender-sensitive, women coming under more pressure to provide good meals with less food, and stressed out by the need to address their children's hunger. These stresses propel women into poorly paid informal sector work, competing among themselves for ever more inadequate earnings. Men also feel the effects, the food price rises severely undercutting their capacities to provide for their families, leading to arguments in the household, fueling alcohol abuse and domestic violence. In the worst instances, couples split and women look for better-off partners. Those affected by economic shocks also respond politically, contesting official explanations of the causes of price hikes, and criticizing their governments for failing to act effectively, blaming corruption and collusion among politicians and business interests.

In one Bangladeshi village visited by the Oxfam researchers, some people were believed to be accumulating microcredit loans simply in order to make their existing loan repayments; default rates were believed to have risen. Many people spent less on personal items like clothes and cosmetics, and scaled down their social lives. While government safety nets provided some support, this had generally failed to protect people from the effects of the price rises. The result of these adjustments was generally not starvation, but an increased level of discontent and stress.

## 4.2 HUMANITARIAN AID

Humanitarian assistance refers to the aid and action designed to save lives, alleviate suffering and salvage some human dignity, following natural or man-made calamities which compromise food and nutrition security *inter alia*. It is distinguished from other forms of aid by its impartiality, its beneficiaries being targeted solely on the basis of need. Food *per se* is an important component of humanitarian aid, though not of course the only one.

Much international government funding is routed through humanitarian assistance expenditure of the 29 members of the Development Assistance Committee (DAC) of the OECD—Australia, Austria, Belgium, Canada, Czech Republic, Denmark, Finland, France, Germany, Greece, Iceland, Ireland, Italy, Japan, Korea, Luxembourg, the Netherlands, New Zealand, Norway, Poland, Portugal, Slovak Republic, Slovenia, Spain, Sweden, Switzerland, the United Kingdom, the United States and European Union[1] institutions—a subset of Overseas Development Assistance (ODA), which is reported to the OECD DAC each year. There is also expenditure by governments outside of the OECD DAC, as reported by the Financial Tracking Service of the UN Office for the Coordination of Humanitarian Affairs (OCHA). Additionally, there are contributions from private sources (including NGOs, trusts, corporations, foundations and individuals).

OCHA is responsible for coordinating responses to emergencies, which it does through the Inter-Agency Standing Committee, members of which include the UN system entities most responsible for providing emergency relief (UNDP, WFP, UNHCR, UNRWA and UNICEF). The UN Central Emergency Response Fund (CERF), managed by OCHA, receives voluntary contributions year-round to provide immediate funding for life-saving humanitarian action anywhere in the world, and the FAO Global Information and Early Warning System (GIEWS) issues monthly reports on the world food situation, with special alerts to Governments and relief organizations identifying countries threatened by food shortages.

Absolute amounts of aid, in cash or kind, channeled for such purpose is difficult to quantify with certainty, as the sources are many. A trusted independent body which collects and collates such information is Global Humanitarian Assistance (GHA), a program run by the Development Initiatives organization and funded by the governments of Canada, the Netherlands, Sweden and the United Kingdom.[2]

The global amount required for humanitarian aid is based on that requested through UN-coordinated appeals each year. In most cases, the initially recorded requirements are revised upward during the course of the year. However, as reported by GHA, in some cases, such as Somalia, South Sudan, Kenya and Yemen in 2013, the original requirements were higher and appeals were actually revised downward during that year.

How much is spent on humanitarian assistance? Is it enough? Where does it go? How does it get there and what is it spent on? Knowing who is

---

[1] Through ECHO (European Commission's Humanitarian Aid and Civil Protection department), more than one-third of the EU annual humanitarian aid budget is used to provide emergency food and nutrition assistance (through food, cash or vouchers), making it one of the world's major donors of humanitarian food assistance.

[2] Retrieved from www.globalhumanitarianassistance.org (accessed 01.04.15).

spending what, where and how is necessary to ensure that resources can best meet the needs of people living in crises, and GHA has produced an annual report since 2000 endeavoring to address such. Its 2014 report looked back on the previous "extreme" year, in terms of both the scale of high-level crises and levels of response. 2013 was marked by high-profile crises in Syria, the Philippines and the Central African Republic, as well as high levels of need both on and off the international radar including in South Sudan, Yemen and the Sahel. As a barometer of global humanitarian need, UN-coordinated appeals targeted 78 million people for assistance in 2013 and called for US $13.2 billion in funding. That year the number of internally displaced people reached an unprecedented 33.3 million, while the number of refugees increased to 16.7 million. Overall, the 2013 appeals were 65% funded. This was the highest proportion since 2009, yet it still left over one-third of identified humanitarian needs unmet.

The GHA report of 2014 records that the level of international humanitarian response rose to a record US$22 billion in 2013. Government donors accounted for three quarters of this, contributing US$16.4 billion—a rise of 24% from 2012, with 9 of the 10 largest government donors increasing their funding. The role of governments outside the OECD DAC has also continued to rise, with this group contributing US$2.3 billion in humanitarian assistance in 2013, representing 14% of the total from all government donors. Private sources provided an estimated US$5.6 billion—a 35% increase from 2012.

Seventy-eight percent of humanitarian spending from OECD DAC donors went to protracted emergencies in long- and medium-term recipient countries. Even before the 2013 escalation in the crisis, Syria received by far the largest volumes of humanitarian assistance: in 2012 it received US$1.5 billion—almost double the US$865 million for South Sudan, the next largest recipient. Funding priorities, political factors and public profile create an uneven global distribution of assistance. Afghanistan, Somalia, Sudan, Ethiopia and the Palestinian West Bank & Gaza Strip have consistently featured in the top 10 recipients list over the past five years. Despite the widely recognized importance of national and local NGOs in humanitarian preparedness and response, they directly accessed only US$49 million of international humanitarian assistance in 2013, a decrease of US$2 million from 2012.

Timely response is critical for effective humanitarian action but, even for acute crises triggered by sudden natural disasters, it inevitably takes time for donors to respond at scale. For instance, in 2013 Typhoon Haiyan (known locally as Yolanda) hit eastern Samar Island of the Philippines on November 8 at 4.40 am local time, and was the deadliest in the country's recorded history. The response to the UN-coordinated appeal during its first month was half that of the Indian Ocean earthquake-tsunami appeal in 2005 in terms of needs met. By their nature, conflict-related crises attract even slower responses.

Humanitarian assistance is not limited to a short emergency phase. Protracted crises continue to capture the bulk of official humanitarian assistance—66% in 2012—highlighting the need for both multi-year funding and better links with development spending and other resources. In most countries, the domestic response to crises goes unreported to international systems. As a result, there is no reliable global figure for this critical and primary response, though data are available for some of the laudable organized national responses. For instance, national budgets show that between 2009 and 2012, India's domestic government resources for disaster relief and risk reduction amounted to a massive US$7 billion, compared with the US$137 million which India received in international humanitarian assistance. The government of the Philippines has similarly and consistently eclipsed international contributions and, in response to Typhoon Haiyan, also demonstrated the primary coordinating role a domestic government can play in disaster relief.

With domestic government expenditure across developing countries now exceeding US$6 trillion a year, these resources can support people's long-term resilience to shocks. But for many countries, particularly those facing entrenched crises, *per capita* spending by those national governments remains low with little prospect of growth. There were an estimated 179.5 million people living in extreme poverty in countries classified as receiving long-term humanitarian assistance in 2012. Almost 40% of long-term humanitarian assistance went to countries with government expenditure of less than US $500 per person per year—one quarter of the developing country average. Where governments lack the capacity or the will to address the risks and needs faced by the most vulnerable people, international resources continue to play an important role. GHA points out that as those worst affected by humanitarian crises are also the most vulnerable—people facing poverty, insecurity and marginalization—it is vital that all resources (public, private, domestic and international) are used coherently.

## 4.3 FOOD-FOR-WORK AND CASH-FOR-WORK SCHEMES

There are remedial programs to help address poverty and disasters, called variously Food-for-Work (FFW), Cash-for-Work (CFW) or Food-for-Assets. All three categories fall on the continuum between unsustainable humanitarian aid and development initiatives. All three help restore livelihoods by repairing damage to the working and living environment, helping them achieve their full potential. The programs do this in ways which are appropriate to the situation in hand, through repairing roads, building irrigation canals and community granaries, sculpting terraces on hillsides, planting trees and other such interventions.

FFW involves rewarding public works through payment in kind (food), directly addressing the problem of undernutrition by raising calorie and (hopefully) significant protein intake. FFW is designed to provide short-term relief assistance, with the outcome of supporting the recovery process. The concept is particularly valuable in areas where food availability is low and markets dysfunctional, and where theft and corruption are major issues.

A downside of this scheme is that food is clearly expensive to procure, store safely and transport to site. The World Food Program has much experience of FFW, distributing food while simultaneously promoting agricultural production and economic and environmental stability; for instance, Bangladesh has built enormous experience in its dry season FFW program, started in 1975 (Ahmed et al., 1995). The FFW program implemented in India by the Catholic Relief Services in the 1980s, helped smallholders below the poverty line who had less than 5 acres, resulting in a threefold increase in cropped area, with agricultural output and household income raised between 39% and 70% (Bryson et al., 1991).

In India, the *National Food for Work Program* was launched in November 2004 in 150 of the most underdeveloped districts, with the objective of generating supplementary wage employment, securing food security and building rural assets. Under the Ministry of Rural Development, the program is open to all rural poor prepared to undertake manual unskilled labor. Implemented as a centrally sponsored scheme, food grains are provided to the States free of charge. The transportation cost, handling charges and taxes on food grains are, however, the responsibility of the States.

Nevertheless, food can be a controversial form of food assistance (Shaw, 1995). While food aid saves lives in emergencies, increases equity and helps communities and countries stabilize economic growth, it has its detractors. They point to the political and commercial motives that sustain food aid flows, its potential disincentive to local agriculture and the risk of increasing dependence on imported foods. Most food aid does not address emergencies, but comprises program aid to countries to help governments with balance of payments issues.

For instance, Ethiopia has been reliant on food aid to make up its national deficit. The fall of the military regime in 1991 and establishment of the Transitional Government of Ethiopia brought a policy shift that signaled a move away from food handouts: the National Disaster Prevention and Preparedness Strategy of October 1993 stated that no able-bodied person should receive gratuitous relief. This led to a guiding principle that 80% of food aid should be distributed through FFW. The official emphasis on FFW gave further impetus to both new and existing projects which stemmed from previous famine relief efforts in the country over the previous twenty years. A range of organizations is involved in funding and implementing such schemes, including WFP, various nongovernmental organizations and the government itself.

By contrast, CFW schemes provide food indirectly, through rewarding individuals for work with cash which they can spend in the ways they choose, including more and/or better-quality food. This scheme works well where local food supplies are available, but people too poor to economically access them. CFW is also better able to stimulate farming activities and local markets where they are present, and also does not cause distortion of prices in the local market through flooding it with food from external sources. As for FFW, the work involved comprises public works to benefit the community, an example being the PSNP in Ethiopia, handled by the Government and the World Bank Trust Fund (Case Study 1 on the book's companion website).

Each category of intervention has an advantage over direct food aid as the output of the work should have a beneficial outcome on local food production and economies, and strengthen long-term food security by improving local infrastructure and/or agricultural potential. Also, both interventions reduce vulnerability to natural disasters and food insecurity by providing a social safety net. There are multiplier effects too, shown in the Food and Cash Transfer program in Malawi, for instance, in which each dollar transferred through the program had economic effects more than double the original input (Magen et al., 2009). As well as governments, NGOs can and have sponsored such programs too, supported by donors such as the Swiss Agency for Development and Cooperation (SDC).

Employment schemes to improve food and nutrition security are well presented in a book by the International Food Policy Research Institute (IFPRI, 1995), with chapters relating to Latin America, China, India, Botswana, Tanzania, Niger, Zimbabwe and Ethiopia.

WFP's *Food Assistance for Assets (FFA)* program uses a term which specifies the output rather than the type of input, helping meet the immediate food needs of vulnerable people while having them build or boost assets that will benefit the whole community. Together this helps make individuals and communities more resilient. The FFA program distributes food itself, food vouchers or cash transfers, so is a composite of FFW and CFW programs.

Helping ensure the poor have access to food even during a crisis and addressing the root causes of food insecurity and vulnerability, FFA programs make important contributions to the second pillar (100% access to adequate food all year round) of the UN's "Zero Hunger Initiative" while contributing to the other four pillars.

All three schemes referred to have little to offer directly to those who are unable to carry out physical work, due to infirmity or because their social obligations like looking after children are too onerous. An exception could be that infirm individuals could participate in teaching programs to build the capacity of the community, mentoring them in long-lost skills for instance. Also, those who are willing though unable to participate benefit indirectly by improvements to watershed management, and the rehabilitation of roads. Humanitarian aid and social safety nets need to be there for those who are unable to provide work for cash or food.

To render the schemes equitable, there has to be a decentralized targeting of beneficiaries and planning. Community-elected committees need to be in place or established to engage in transparent and accountable community-level targeting of the poor, and to achieve active local participation and leadership in project identification, design and implementation of community works. This is the case in Ethiopia, for the PSNP.

## 4.4 EU FOOD FACILITY PROGRAM

### 4.4.1 Background

The volatility of food prices, agricultural inputs, and fuel in 2007 and 2008 put numerous developing countries and their populations in an existential quandary (see Section 3.4 and Case Study 9 on the book's companion website), and the realization of Millennium Development Goal 1 at risk (Case Study 8 on the book's companion website). Though the rise in food prices since 2006 has been fairly general, the increase in volatility is confined to grains and some vegetable oils, which are the main determinants of food availability and access. The 2008 crisis resulted in riots, unrest and instability in many countries (such as Egypt, Cameroon, Ivory Coast, Burkina Faso, Senegal, Madagascar, Mozambique, Somalia, Philippines, Bangladesh and Haiti), jeopardizing the achievements of years of political development and peacekeeping investments. This FPC required short-medium term action from the international community, in order to mitigate its effects on the poor in developing countries.

FAO launched its Initiative on Soaring Food Prices (ISFP) in December 2007, and in April 2008 the Chief Executives Board of the United Nations established a High-Level Task Force (UNHLTF) on the Global Food Crisis, under the leadership of the UN Secretary-General. This Task Force proposed a unified response to the FPC and a global strategy and action plan, the so-called Comprehensive Framework for Action.

The FAO High-Level Conference on World Food Security in June 2008 called on the international community to increase its assistance to developing countries. This call was echoed by the G8 + Summit in Japan the following month, at which the President of the European Commission announced the Commission's intent to propose a rapid response "Facility" of 1 billion EUR to mitigate the effects of the ongoing FPC.

The first EU intervention decisions made in 2008 to mitigate the FPC originated from existing EC instruments in support of food security: the EDF-B envelope (185.9 million EUR), Food Security Thematic Program (FSTP) (50 million EUR) and ECHO (210 million EUR). However, the amount made available or reallocated through these existing instruments was insufficient to meet the financial need for responding to the FPC. On December 16 2008, the European Parliament and the Council adopted Regulation 1337/2008, establishing a "facility for rapid response to soaring food prices in developing countries" (which was to become the European Union Food Facility—EU FF). Operating over a 3-year period from 2009 to 2011, the fund was intended to bridge the gap between emergency aid and medium- to long-term development aid.

The Facility's primary objectives were to: (1) encourage food producers to increase *supply* in targeted countries and regions; (2) support activities to respond rapidly and directly to mitigate the negative effects of volatile food prices on local populations in line with global food security objectives, including UN standards for nutritional requirements; and (3) strengthen the productive capacities and governance of the agricultural sector so as to enhance the sustainability of interventions.

### 4.4.2 Implementation and Evaluation

Fifty-seven percent of the total EU FF budget was channeled through UN agencies and the World Bank, while the remainder was disbursed through Budget Support, non-State actors, member state agencies and regional organizations. The Food Facility was the largest additional global grant contribution to stimulate agricultural development and promote food security since the G8 leaders pledged their support in this respect at the 34th Summit at Toyako, Japan in July 2008.

An external Final Evaluation[3] was mounted by the EC of the instrument itself, and the Commission's cooperation activities under it over the period 2008—2011, across all 49 countries where EU FF activities were undertaken. The objective of the evaluation was to ascertain whether the objectives had been met, to enable the formulation of conclusions based on objective, credible, reliable and valid findings, and to formulate recommendations with a view to improving relevant future development cooperation operations. Based on a desk study of all 232 interventions under the EU FF, and field visits to 12 of the 49 beneficiary countries,[4] some major conclusions reached on the EU FF Instrument and its execution were as follows:

1. Adverse external factors, the continuous volatility of food price and the short implementation period of the EU FF, meant that conferred resilience to food insecurity in the countries targeted was incomplete. There was little evidence of the major effects of the EU FF on food prices or food security beyond the direct beneficiaries.
2. The FPC was seen primarily as an economic crisis. The solutions that were applied were primarily traditional food security interventions. There was a predominance of interventions related to agricultural and local development, but only a few interventions in favor of urban populations or improving nutrition.
3. Most projects financed under the EU FF were targeted at production while only a minority introduced safety net measures. Most interventions in rural areas targeted poverty reduction and local development rather than focusing on "mitigating the effects of the FPC". Beneficiaries were poor and often insecure farmers, whereas landless farmers and urban poor were insufficiently targeted.

---

[3] Retrieved from http://ec.europa.eu/europeaid/what/development-policies/intervention-areas/ruraldev/documents/euff-final_report-en.pdf (accessed 02.03.15) (quotations from this with permission from the Publications Office of the European Union).

[4] A sample of 12 countries was selected according to a set of criteria defined in the EU Food Facility final evaluation Inception Report.

4. In most cases there was no strategy to replicate experiences (of EU FF interventions) in other parts of partner countries (with other external funding or national funding). The sustainability of most activities was not assured, this depending on whether the national Government, the EU or other donors would continue to support beneficiaries in order to consolidate achievements.
5. Food prices were higher at the end of EU FF implementation than at the beginning. Already in 2008, international organizations indicated that the market for food crops would remain volatile in the coming years. This has proven to be true, as, after a strong decrease of food prices in the second half of 2008, there was an upward trend in 2009, with a peak reached in 2011. Many countries supported under the EU FF continue to be food-insecure.
6. The EC, through its decision to support the coordinated international response to the FPC, brought food security and rural development to the forefront of its own development cooperation agenda and of the international development agenda.[5] The EU FF enabled rapid increase in the overall volume of funds directed through EU cooperation to the agricultural and rural sector. Although the EU re-deployed other instruments in response to the crisis, these were not sufficiently flexible or lacked sufficient funding to allow the necessary response to the crisis. So, the creation of a specific instrument had been necessary.
7. The EU FF was fully coherent with both the Paris Declaration and the Declaration of the World Summit on Food Security, and supports the four pillars of Food Security in countries affected by the FPC. However, support was concentrated on triggering a supply response from the smallholder farming sector.
8. There was a broad consistency and coherence between EU FF interventions, other EC instruments and other donors' interventions. Coordination was also satisfactory.
9. The EU FF was programmed as a short to medium-term instrument and its activities were concentrated in short-term support in order to act as a bridge between emergency and long-term development, even though by 2008 it was expected that food prices would remain volatile in the mid-term. Due to the short time period available, interventions concentrated on promoting resilience and reducing the effects of the FPC on the most affected groups. However, the underlying causes (both natural and man-made) of food insecurity in target countries were out of the scope of the EU FF. Although there were no plans for continuing the EU FF with another instrument at the time of the final evaluation, a number of projects financed under the EU FF were foreseen as benefiting from further financing under long-term instruments (FSTP, EDF).

[5] The share of Overseas Development Assistance devoted to agriculture reached a level of 19% in 1980, but fell to 3.8% in 2006. However, it seems that this trend is reversing.

10. The initial programming of interventions was carried out on the basis of Country Needs Assessments (CNAs) performed by UN organizations. These CNAs revealed the extent of the crisis and proposed responses in each country. However, CNAs were somewhat limited by lack of involvement with other key stakeholders and by not considering Non-State Actors (NSAs) and Budget Support (BS) as possible channels of support.
11. Interventions financed under the EU FF were effective in both mitigating the effects of the FPC on the direct beneficiaries of the interventions, and in promoting resilience within these populations. The FPC has caused or aggravated food insecurity in target countries. Although the support provided under the EU FF arrived some months after the peak in global food prices, it did correspond to the needs of the beneficiaries, and arrived in time to deal with the effects of this crisis and tackle food insecurity.
12. Projects tended to be overambitious in their design given the time available.
13. All interventions were in line with, or at least were not counter to, national policies for the agricultural and food security sectors.

The Final Evaluation provided several *recommendations*, including:

- The EU should consider *converting the EU FF into a permanent "Stand-by" instrument*, in order to respond rapidly to future sudden Food Price Crises. *Food Security should remain at the top of long-term programmed cooperation* in line with the *Agenda for Change*.
- The design of future specific instruments should be *more focused*, concentrating support on the most affected countries to ensure resources are allocated for maximum impact.
- When selecting partners and aid modalities, the EU should exploit the *comparative advantages* provided by the various implementation channels.
- The EC should *continue to play an active role in policy dialogue* at country level, stressing the importance and multisector dimension of food security.
- EU interventions should endeavor to systematise *lessons learnt*, and *share experiences* with implementing partners. Specific attention should be given to *replicability and cost-benefit analysis* of interventions.

A strategic lesson learnt, as pointed out under the Final Evaluation, was that an instrument that is designed to respond to a single challenge (the food price crisis) must take into account in its programming the many other external factors that affect food prices and availability. In this case, the EU FF had to work against a background of climate change, global fuel price and financial crises and many other regional and national crises such as droughts, floods and earthquakes.

The European Court of Auditors (2012), in a parallel and independent report published in the same year made a similar assessment of EU development aid in support of food security, specifically for sub-Saharan Africa. Recognizing that in sub-Saharan Africa, 30% of the population suffers from hunger, the Court report concluded that EU development aid channeled towards enhancing food security (availability and access) in sub-Saharan Africa is largely effective and relevant to the countries' needs and priorities. Moreover, that the interventions are mostly well-designed, achieve most of their expected results and that half of them are likely to be sustainable. However, the Court recommended that to improve the effectiveness of EU development aid for food security in sub-Saharan Africa, the Commission should:

- examine the feasibility of a permanent instrument for financing urgent and supplementary measures to address the consequences of potential future food crises in developing countries,
- increase priority for nutrition when defining cooperation strategy, and encourage countries to devise appropriate nutrition policies and programs,
- set out more precise intervention objectives, measurable through performance indicators.

## 4.5 OTHER EFFORTS BY DONOR PROGRAMS TO IMPROVE AGRICULTURAL PRODUCTIVITY AND MARKETING

There needs to be an integrated approach to tackling the root causes of food insecurity. In terms of *food availability*, this includes achieving higher productivity and diversification of food production, which may be constrained by environmental degradation and climate change. Human *access to food products* requires functional markets and safety nets, incremental income generation, insurance schemes, farmers' access to markets and financial services, and emergency stocks. It also requires investments in agribusiness, distribution systems and other rural infrastructure, research and technology transfer. From a long-term perspective, support to sustainable agriculture is paramount to build up resilience (see Section 5.3) in sub-Saharan Africa, for example, where the sector provides employment to 60% of the population, including the most vulnerable. Some examples of donor partner-funded programs, addressing one or more of these themes, are given in the following sections.

### 4.5.1 Australian Government and FAO Support to the Philippines

The Australian Government's strategy for the rural development sector through its AusAID-DFAT (Department of Foreign Affairs and Trade, Australia) program is to focus on reducing rural poverty, by increasing opportunities for the poor to generate income. Rural development assistance is provided for agriculture, fisheries, forestry and research. The three components of the overall program are: increasing agricultural sector productivity, stimulating rural nonfarm employment and managing natural resources sustainably. Australia takes a three-pronged approach to working with partner countries:

- assisting partner governments to develop and administer policies that will promote income generation;
- working directly with rural communities on income generating projects; and
- developing collaborative partnerships in agricultural research for development.

In the Philippines, several projects have been implemented over the years in support of Australia's Development Cooperation Program strategy. The Philippines-Australia Community Assistance Program (PACAP) supports community-based projects through directly funding Philippine NGOs and People's Organizations. Since October 2011, Australia has committed about Php 110 million (A$2.5 million) grant funds for 62 community-based projects spread across Luzon, Visayas and Mindanao. Of these 62 projects, 29 and 33 are ongoing and completed, respectively.[6] Typhoon Yolanda struck the Philippines in late 2013, and survivors are still reeling from the severity of the damage brought by the tropical cyclone. PACAP is helping the Philippine government restore livelihoods and rebuild communities in the Visayas region which suffered the brunt of the storm.

In 1999, the 10-year Philippines-Australia Local Sustainability (PALS) Project commenced implementation to assist Local Government Units (LGUs) and local communities plan and manage sustainable activities to improve the livelihoods of the rural poor in the province of Misamis Occidental. These include the development and implementation by targeted local communities of microprojects, including those in agriculture, fisheries and forestry. Examples of local government infrastructural activities included farm-to-market roads and farm mechanization.

The Australian government has also collaborated since 1996 with FAO on a Foot and Mouth Disease (FMD) Project to enable the Southern Philippines to be declared free of the disease. Furthermore, the Philippines-Australia Technical Support for Agrarian Reform and Rural Development (PATSARRD) project will be executed by FAO, and is expected to commence in 2016, helping beneficiary families improve their economic and social conditions.

### 4.5.2 USAID Support to Afghanistan's Agriculture

Agriculture is critical to Afghanistan's food security and a key driver of economic growth. Sixty percent of Afghans rely on agriculture for their livelihoods. The sector accounts for about 40% of Afghanistan's GDP. Prior to decades of conflict, Afghanistan's agricultural products earned a global

[6] Retrieved from http://pacap.org.ph/ (accessed 11.05.15).

reputation for excellence, particularly almonds, pomegranates, pistachios, raisins and apricots. Decades of war and neglect have devastated Afghanistan's farmland, displaced millions of people and largely destroyed the country's existing infrastructure.

Currently, there are 11 ongoing USAID agricultural support projects in Afghanistan.[7] US assistance to Afghanistan's agricultural sector focuses on creating jobs, increasing incomes and productivity, enhancing food security, creating export markets and strengthening the Afghan government's ability to promote broad-based growth. Since 2002, USAID, together with the United States Department of Agriculture (USDA) and the Afghan government, has:

- facilitated over $306 million worth of increased sales of licit farm and non-farm products;
- created over 358,000 jobs through licit livelihoods activities;
- brought more than 1 million hectares under improved management, through tree planting and improved water management practices;
- rehabilitated irrigation infrastructure over 106,000 ha to enable multiple cropping;
- supported every link in the value chain—market-led production and growth, storage, processing, shipping and sales. USAID has facilitated agricultural export sales of fresh and dried fruit and nuts and cashmere worth $54 million, to India, Pakistan, UAE, UK and other countries;
- USAID has helped 1.1 million households, planted over 3.9 million fruit saplings and grape cuttings, established over 25,000 hectares of orchards and vineyards, and built 200 raisin drying facilities and storage rooms;
- through improving farmer's access to credit using the Agriculture Development Fund, which it helped create, USAID has provided $53 million in loans to farming households and agribusinesses.

### 4.5.3 USAID Support to Liberian Agriculture

One of the recent USAID-funded projects in Liberia has been the Economic Growth Corridor Study, Phase 2, in 2011. The Ministry of Planning and Economic Affairs was the host government agency. Phase 2 of the Study addressed two of the five key growth corridors (Monrovia-Ganta and Buchanan-Yekepa corridors) identified in Phase 1, in 2010. The Monrovia-Ganta corridor comprises the backbone of the Liberian economy, having some 43% of the population. Both corridors potentially link the interior of Guinea and Ivory Coast to major Liberian ports.

The study team comprised international specialists in agriculture, trade and investment, business and regional planning, together with a team of Liberian analysts. Using a combination of Participatory Rural Appraisal (PRA) questionnaires, focus group discussions and physical inspection,

---

[7] Retrieved from www.usaid.gov/afghanistan/agriculture (accessed 10.05.15).

quantitative and qualitative data were collected to develop a framework for a comprehensive value-adding growth corridor strategy to underpin Liberia's Economic Growth & Development Strategy, and augment the Lift Liberia Poverty Reduction Strategy.

Eleven agricultural investment proposals were prepared, at prefeasibility level (with gross margin analyses), designed to attract private and corporate investors. The value-adding projects involved cocoa, oilpalm,[8] vegetables, vanilla, agroindustry, ecotourism, rural markets, warehousing and a road transport system. Support to smallholder cash crop farmers' organizations was a major substrategy espoused in two of the proposals, to reduce transaction costs and foster confidence in advancing from smallholder farming to semi-commercial, which would transform the food security status of the country. The first project to be subsequently selected and implemented was the upgrading of the marketplace in Ganta, Nimba county, the second largest city in the country lying on the border with Guinea.

### 4.5.4 African Development Bank, Japanese Government and UNDP Support in Africa

By the mid-1990s, rice production in sub-Saharan Africa was being outstripped by rapid population growth. The resulting rice imports to feed 240 million people in West Africa alone were taxing foreign reserves by nearly US$1 billion annually. Worse still, most rice growers were facing the unenviable choice between a high-yield species poorly adapted to African conditions (the Asian rice *Oryza sativa*, adapted for irrigation) and a well-adapted but low-yielding species (the African rice *Oryza glaberrima*, adapted to dryland conditions).

NERICA (New Rice for Africa) was launched in 1996—an interspecific hybrid between the Asian and African rice—a high-yielding, drought-resistant, and protein-rich variety that has contributed to food security and improved nutrition in several countries in western and eastern Africa, including Congo Brazzaville, Cote d'Ivoire, the Democratic Republic of the Congo, Guinea, Kenya, Mali, Nigeria, Togo and Uganda. NERICA rice was developed at the Africa Rice Center (formerly WARDA until 2009), funded jointly by the African Development Bank, Japanese government and UNDP. Since 1996, some 18 cultivars of the hybrid species have been made available

[8] In the case of oilpalm, a huge upscaling was suggested in using the hand-cranked pressing machine developed under another USAID-funded project in Liberia, implemented by the NGO Winrock. This saves so much physical labor and time for rural women, is far more hygienic than the traditional way, produces a better quality oil and far more of it from a given weight of fruit, and provides employment for local artisans in making the presses from recycled scrap metal—a comprehensive win-win for everybody concerned.

to rice farmers across sub-Saharan Africa, through the agency of a *powerful coalition of governments, research institutes, private seed companies and donors*. For the first time, many farmers have been able to produce enough rice to feed their families and make a profit in the market.

For their work in developing NERICA, two scientists, working independently, were awarded the annual World Food Prize in 2004. These were Professor Yuan Longping, Director General of the China National Hybrid Rice Research and Development Center in Hunan, China, and Dr Monty Jones of Sierra Leone, a former senior rice breeder at the West Africa Rice Development Association.

### 4.5.5 UNDP and USAID Support to Jamaica

Jamaica is on track to eliminate hunger, though the economic and food crises of recent years may jeopardize that goal. The percentage of the population living in poverty dropped from 28.4% in 1990 to 9.9% in 2007, but rose to 12.3% in 2008, 16.5% in 2009 and 17.6% in 2010.[9] A one million dollar Rural Youth Employment project was conducted from 2010 to 2013 to improve employability in the farming and agroprocessing sector. Nearly one-third of Jamaicans aged 15–29 are unemployed, and around 650 of whom, men and women, from four impoverished areas of Jamaica, were beneficiaries of this project. Additionally, more than 360 high-school students participated in agriculture career days.

The project was implemented by Jamaica's Scientific Research Council in partnership with the National Center for Youth Development, the Rural Agricultural Development Authority and national youth organizations, and funded by UNDP and USAID. The project provided not only technical knowledge and training, but facilities and equipment that the trainees would not otherwise have been able to access. The project also involved workshops and career days, with presentations by farmers and agribusiness professionals from around the country. The trainees were thereby given a wide range of experience and network of contacts which they could use to secure jobs, or finance for their own enterprises following their training.

## 4.6 ENHANCING THE RESPONSE TO CRISIS

The EC in its Communication COM 586 of 2012 (EC, 2012) offers valuable insights into ways to improve the impact of donor partner responses to crises, in particular the EC's own response (see Box 4.1).

---

[9] Retrieved from www.latinamerica.undp.org/content/rblac/en/home/ourwork/povertyreduction/successstories/jamaica-agriculture.html (accessed 10.05.15).

### Box 4.1 Enhancing the Response to Crises

*...the following elements can help improve the impact of the responses to crises when they strike:*

1. *The* ***preparation of a joint analytical framework*** *prepared by both humanitarian and development actors that:*
   - *identifies the root causes of the crisis as well as the precise impact on the most affected populations.*
   - *assesses on-going interventions to see if the root causes are being addressed, and also to see if there are gaps in the assistance that is being provided.*
   - *identifies the areas, both in terms of sectors and geographic regions, where an enhanced resilience approach could have the most impact.*
   - *defines strategic priorities for the short-term (early recovery) as well as for the long-term, within a coherent "Resilience Approach"*
2. *There needs to be an* ***increase in short-term financing to support the early recovery phase****. Recent initiatives highlight the need for a higher degree of flexibility in programming to react to fast changing needs, without reducing on-going medium/long term activities to address root causes. New modalities of assistance like EU Trust Funds should be considered to tackle emergency or post-emergency situations.*
3. *Most major crises span across borders. The* ***capacity of regional organizations*** *needs to be strengthened so that they can develop cross-border initiatives and promote regional integration.*
4. *For major crises, light structures should be set up to enable* ***donor coordination and a structured dialogue established with partner countries and regional organizations****. It is necessary to define and formalise who does what, based on the comparative advantage of each actor in a given context. Both development and humanitarian actors should be actively engaged.*
5. ***Finding short-term interventions that have a long term impact.*** *Even though short-term responses, and humanitarian assistance in particular, are mostly focused on life saving and asset protection, such activities can also have a long-term impact. For example, shifting from food aid to cash transfers can have a long-term effect in stimulating the local market and financing public works that can reduce the likelihood of future disasters or mitigate their impact. These types of intervention should be identified and prioritised.*
6. *Where violent conflicts exist, the resilience strategy and the wider EU political and security approach should be mutually supportive and consistent, and synergies should be developed at the levels of instruments, notably the Common Security and Defence Policy instruments and the Instrument for Stability.*

**Source: From Section 4.3 of COM 586 (http://eur-lex.europa.eu/procedure/EN/202008, quoted with permission)(bold typeface and numbering added) "http://eur-lex.europa.eu, © European Union, 1998–2015".**

## 4.7 ADDRESSING NUTRITION INSECURITY

### 4.7.1 Interventions to Address Protein-Calorie Undernutrition

Bergeron and Castleman (2012) *op. cit.* explain that the initial intervention model to combat *acute malnutrition* was community therapeutic care, which focused on emergency settings. Adopting the more recent decentralization of management, this model evolved into the Community-Based Management of Acute Malnutrition (CMAM) model, now the global standard, which incorporates treatment of acute undernutrition both in development settings and in routine health services. The effectiveness of CMAM relies on an enabling policy environment; adequately trained personnel; an effective supply chain management system for medications; adequate infrastructure for screening, inpatient and outpatient care; effective community outreach and active case finding; functional follow-up and referral systems; and, last but not least, active community participation. Overall management should ideally be hosted in individual Ministries of Health, with technical assistance from specialized agencies such as UNICEF and specialized NGOs, for technical backup and capacity building of local staff, and financial assistance from the UN or other donor to cover logistics and other needs.

By contrast, long-term *chronic malnutrition* in children is addressed through focus on maternal nutrition during pregnancy and lactation, and protecting the health and nutrition of the child during the first 2 years of life. Community-based approaches have been developed covering health, diet and care. This model emphasizes addressing the underlying food security determinants of nutrition—such as income, and access to food/water/sanitation/health services and education/women's empowerment. The compound multisectoral causes of food insecurity as expressed by chronic undernutrition demand a comprehensive and coherent approach to mitigating the situation. Thus, a fusion is needed to include *inter alia* gender-sensitive social protection measures, development initiatives to improve agricultural productivity, water and sanitation and counseling to provide advice and support to mothers.

### 4.7.2 Interventions to Address Micronutrient Deficiencies

The most effective way to meet community health needs safely is by population-based approaches involving *food fortification*, which has a century-long record of success and safety (Tulchinsky, 2010) *op. cit.* WHO defines fortification as the practice of deliberately increasing the content of an essential micronutrient in a food, in order to improve the nutritional quality of the food supply and provide a public health benefit with minimum health risk. Wheat flour is the most consumed cereal flour in the world, and for more than 60 years its fortification with vitamins and minerals has been effective, especially for folic acid, iron, zinc and other B vitamins. Chile was the first of many developing countries to follow the North American example of mandatory fortification of flour with folic acid. Flour fortification is a simple technique, requiring only slight modifications in most flour mills. Other foods can also be fortified, such as rice (in Asia, the Pacific, Latin America, and the Caribbean), sugar (as in Brazil, El Salvador, Guatemala), salt (India), fish sauce (most suitable in Vietnam to reach maximum numbers of people), soy sauce (in China, where it is consumed by 80% of the population), and edible oil (this host commodity having a large potential for West Africa).

Another strategy is the use of *dietary supplements*. A dietary supplement is a product intended for ingestion that contains a "dietary ingredient" intended to add further nutritional value to the diet. A "dietary ingredient" may be a vitamin, mineral or amino acid, for instance, or a combination of these. Dietary supplements may be found in many forms such as tablets, capsules, softgels, gel caps, liquids or powders/sprinkles. Examples of supplementation initiatives include iodized salt programs, first introduced in Switzerland in 1923, which have led to marked decrease in prevalence of goiter where implemented. Iodized salt is the primary strategy for correcting iodine deficiency due to the near universal consumption of salt regardless of socioeconomic status. Of the 130 or so countries which experience iodine deficiency in their populations, three quarters have laws mandating salt iodization. For instance, in the National Program for Food Security of the government of Mongolia (2009–16) it is stated that 96.3% of salt used in the country is iodized (and more than 30% of domestically-produced flour is fortified with iron and vitamins) (FAO and WHO, 2008).

Dietary supplements are normally added to the food just before it is eaten. Micronutrient Initiative et al. (2009) *op. cit.* reports on an example of successful use of dietary supplementation, again in Mongolia, to address the prevalence of anemia in children there aged 6–35 months. World Vision, in collaboration with the Mongolian Ministry of Health and the Hospital for Sick Children in Toronto, Canada, designed a program for the distribution of sachets of iron- and Vitamin D-rich sprinkles for home-based use. The initial distribution program was in 2001–03, as a result of which anemia prevalence was reduced from 42% to 24%. The program has since been scaled up.

A valuable two-page list of recommended priority intervention actions to redress vitamin and mineral deficiencies, is given by Micronutrient Initiative et al. (2009) *op. cit.* These actions target potential implementers—national governments, the food industry and international organizations.

### 4.7.3 The Copenhagen Consensus and the Scaling-Up Nutrition Movement

At the instigation of Danish academic Bjørn Lomborg, a group of 8 economists (including five Nobel Prize-winners) clubbed together as the *Copenhagen Consensus* panel, to deliver their priorities from 10 leading challenges facing the world, for promoting global well-being and guiding development spending. They deliberated six times between 2004 and 2012, assessing learned economic research papers written for them by scholars around the world on the subjects concerned. Their deliberations in 2008 and 2012 prioritized micronutrient interventions as being the most cost-effective option.

The panel selection has been criticized by equally eminent economists, such as Columbia University Professor Jeffrey Sachs, and gripes even from within the Copenhagen Consensus panel, that the prioritization of interventions accorded in the final report misrepresented the findings, due to dubious weighting mechanisms employed. Nevertheless, that micronutrient interventions were considered, and on the basis of cost-effectiveness were highly valued using welfare economics analysis, should be viewed as objective plaudits, even without reference to other potential fields of intervention.

The three authors of the component paper on hunger and malnutrition (Horton et al., 2008) determined five priorities for investment concerning the smartest way to reduce hunger and undernutrition around the globe:

- micronutrient supplementation, providing Vitamin A capsules and zinc supplements for under 2year olds (cost $60.4million annually, with benefits of more than $1 billion annually, thus a benefit-cost ratio—BCR—of 17:1),
- micronutrient fortification, providing iodized salt and iron (cost $286 million annually, with $2.7 billion benefits, thus a BCR of 9.5:1),
- "biofortification",[10] (cost $60 million annually, with benefits of $1 billion, thus a BCR of 16:1) (see Section 5.6),
- deworming of preschoolers (cost of $26.5 million with benefits of $159 million annually, thus a BCR of 6:1),
- community-based nutrition promotion ($789 million cost, yielding $10 billion benefits, thus a BCR of 12.5:1).

These five were ranked among the 30 priority global interventions suggested by the overall consensus panel, respectively as first, third, fifth, sixth and ninth. Only the first two of the five are solely focused of micronutrients, though all five address undernutrition in general. The subsequent Copenhagen Consensus conference in 2012 assigned "bundled nutrient interventions" as deserving the highest priority.

The *SUN (Scaling-up Nutrition) Movement*, explains in its Progress Report of 2010–11 that it started its work in September 2010 and has the backing of the UN Secretary General and UN Standing Committee on Nutrition, together with various donor partner governments and the Gates Foundation. While acknowledging the prioritization to undernutrition given by the Copenhagen Consensus in 2012, the SUN Movement draws attention to the need for a multi-disciplinary approach to relieving the world of undernutrition and hunger. Understanding of the etiology, epidemiology, scientific-nutritional and sociological basis of undernutrition has increased over the decades. SUN believes that there is little now that the world does not know—it has only to share and apply the knowledge, train and deploy the skills, and encourage and fund the necessary actions among all stakeholder groups (public and private sectors, civil society, international organizations and donors) to overcome the causes of undernutrition.

Technical solutions which correct and prevent micronutrient deficiencies, and are transferable/adaptable to location-specific resource-poor developing world contexts, are only part of the solution. How to grow more food, relieve

[10] An example of this is "golden rice", genetically engineered for the first time in 1999 to contain provitamin A which "normal" rice grains do not contain, and hence combat Vitamin A deficiency (www.goldenrice.org). Adapting local high-yielding varieties with this genotypic feature offers a long-term sustainable option to resolve Vitamin A deficiencies, suitable for transfer to resource-poor contexts, and far cheaper than recurrently needed industrial fortification outreach programs run by UNICEF, the Micronutrient Initiative et al.

poverty, prevent infectious diseases or underwrite maternal care are necessary though insufficient conditions to resolve the scourge of hunger and undernutrition. Equally important is an emphasis on preventing undernutrition in the first place, through addressing underlying political, economic and cultural issues—ensuring a good health service, sound labor laws, nutrition-sensitive agribusiness, effective sanitation and clean drinking water—and the need for governments to take ownership of the social and economic imperatives to drive the initiatives through, with appropriate policies, strategies, laws and actions, getting improved nutrition and funds for it to the heart of the agenda for development. In short, a coherent strategic plan is needed, country-wise, regional and global, fully-funded and -actioned. The SUN movement is a worthy partner in driving the need for such an agenda, and has achieved considerable success with a number of governments in the developing world.

### 4.7.4 Government-Owned Social Transfers to Promote Full Nutrition Security in Nigeria

An example of a comprehensive preventative and curative social protection initiative is given below. It involves access to, and utilization of, nutritious food by infants and their mothers, spear-headed by the Nigerian government, with exemplary support by international agency and NGO partners.

The National Primary Healthcare Development Agency (NPHCDA) is an agency of the Federal government of Nigeria, the responsibility of which is to implement health policies. It is a well-structured organization with a presence in all States and Local Government Areas (LGAs). NPHCDA has zonal offices at Enugu, Ibadan, Benin-City, Bauchi, Minna and Kano. As part of its mandate, it is responsible for the implementation of the biannual Mother and Newborn Child Health (MNCH) weeks in Nigeria. The agency has Primary Health Care (PHC) centers in every ward in the LGAs, with good outreach to vulnerable groups in the LGAs. Additionally, there are Ward Health Committees and Ward Development Committees that operate at ward levels. There are also in place community or village volunteers who can reach the most vulnerable members of the communities in wards of LGAs throughout the country. NPHCDA uses this platform to deliver health commodities such as vaccines, and to mobilize for MNCH weeks and health campaign programs.

Nutrition is a major component of the Agency's service delivery. Technical support is provided to States on maternal nutrition, child nutrition, micronutrient deficiency control and nutrition surveillance. The Agency monitors all these activities and has a well-organized reporting system. It also builds the capacities of Health workers at State level on maternal nutrition, nutrition during pregnancy and lactation, and has designed materials used at ante-natal services. Other nutrition-specific activities include growth monitoring, provision of iron and folate supplements for pregnant women, food demonstrations (not carried out as frequently as desirable due to financial constraints), and training on *complementary* feeding and dietary diversification.

The Agency is currently (in 2014) implementing programs at the Fourth level structure, ie, community level. The community level structure is more effective and fosters sustainability, and comprises the Ward Development Committee and Village Development Committee. Trained workers sensitize the community on nutrition and health programs, reporting to the PHC facilities in the wards. These workers also form part of the Infant and Young Child Feeding (IYCF) support group that monitors breast feeding, undertakes counseling work and house-to-house visits to monitor compliance and render help where necessary, and also helps to keep child health record cards.

Another major area of focus for NPHCDA is the micronutrient deficiency control program. This is anchored in the biannual MNCH week. In this program, there is vitamin A *supplementation* for children 6–59 months, deworming of children, distribution of iron/folate tablets to pregnant women and screening for undernutrition using Mid-Upper Arm Circumference (MUAC) tapes. A consequence of this screening is that children found to be severely undernourished are referred to CMAM centers for treatment/supplementary feeding.

Services provided at CMAM centers are free, such as provision of highly nutritious RUTF, training of mothers in basic hygiene (hand washing) and weekly monitoring of infant weight gain. Improved hygiene is much needed—diarrhea in children is regarded as the "norm" in northern Nigeria, and is associated with much reduced absorption of nourishment in the digestive tract. Awareness campaigns are run for the young mothers (often so young they are hardly out of childhood themselves and needing care, while motherhood obliges them to care for their own infants).

Antenatal clinics at the PHCs are used to pass nutrition information to mothers on the need for early initiation of breast feeding, exclusive breast feeding for the first 6 months of life (as per the WHO definition) and preparation of adequate complementary feeding.

The NPHCDA also uses RapidSMS to transmit information from ward level up to the LGA, State, zone and national offices. RapidSMS is also used to monitor compliance. NPHCDA already has a platform (the Galaxy backbone IT network) for payment of the Midwives service scheme, which could be upgraded for further service delivery in terms of cash transfers. The MNCH week also has a mobile network for transmission of information for messages and data. This system could also be of great use in social transfers. It is operated by the Planning and Statistics Department of NPHCDA.

There are some challenges, however, which include religious refusal of some women to give RUTF to their child, the mother sharing the RUTF with other children at home, negligence of mothers to keep to instructions on the practice of exclusive breast feeding[11] even when they are full-time housewives (without the need to leave home for paid employment), and insufficient money

[11] The "norm" in northern Nigeria is that babies should be given water after initial breast feeding to ensure they "stay honest in future life".

to buy foods and equipment for food demonstrations. How to increase the coverage of the latter all-important food demonstrations across northern Nigeria comprises one of many recommendations that a nutrition-sensitive social transfers study team made to the EU Delegation in Abuja in late 2014. Synergy with NPHCDA is provided by other State government agencies through providing free school meals for primary school children, while the State agricultural authorities contribute important nutrition resilience-building initiatives.

For further information on human nutrition, especially for those involved in humanitarian emergency practice and training in the field, the reader is referred to an excellent manual devised by UNICEF called "Harmonized Training Package: resource material for training on Nutrition in Emergencies (the HTP)". This package is available online and comprises five sections: International Humanitarian System and Reform; Basic Concepts in Nutrition and Emergencies; Measuring Undernutrition in individuals; Micronutrients; and, Infant Feeding in Emergencies[12].

Applying the provisions described in this chapter to mitigate *current* food insecurity is clearly pertinent also to preventing *future* food insecurity, as addressed in Chapter 5 below, "Prevention of Future Food Insecurity".

## REFERENCES

Ahmed, A.U., Zohir, S., Kumar, S.K., Chowdhury, O.H., 1995. Bangladesh's Food-For-Work program and alternatives to improve food security. In: von Braun, J. (Ed.), Employment for Poverty Reduction and Food Security. IFPRI, Washington, DC, pp. 46–74. (Chapter 3).

Bergeron, G., Castleman, T., 2012. Program responses to acute and chronic malnutrition: divergences and convergences. Adv. Nutr. 3, 242–249.

Bryson, J.C., Chudy, J.P., Pines, J.M., 1991. Food for Work. A Review of the 1980s With Recommendations for the 1990s. 81 pp. <http://pdf.usaid.gov/pdf_docs/pnabh222.pdf>.

Davies, S., 1993. Are coping strategies a cop out? IDS Bull. 24 (4), 60–72.

Drèze, J., Sen, A., 1989. Hunger and Public Action. Clarendon Press, Oxford.

EC, 2012. Communication from the Commission to the European Parliament and the Council. The EU Approach to Resilience: Learning From Food Security Crises COM/2012/0586 Final. Oct. 3, 2012.

European Court of Auditors, 2012. Effectiveness of European Union development aid for food security in sub-Saharan Africa. Special Report Number 1 (pursuant to Article 287(4), second subparagraph, TFEU) 70 pp. <http://www.europarl.europa.eu/document/activities/cont/201203/20120329ATT42184/20120329ATT42184EN.pdf>.

FAO and WHO, 2008. Joint FAO/WHO Food Standards Program, FAO/WHO Coordinating Committee for Asia. Sixteenth Session, Denpasar, Indonesia, 17–21 November 2008. Nutritional Issues Within the Region, p. 2.

Horton, S., Alderman, H., Rivera, J.A., 2008. Hunger and Malnutrition. Executive Summary. Copenhagen Consensus Center (2008). Copenhagen Business School, Denmark, 6pp.

Hossain, N., Green, D., 2011. Living on a Spike: how is the 2011 food price crisis affecting poor people ? Oxfam Research Report. 48 pp. <https://www.oxfam.org/sites/www.oxfam.org/files/file_attachments/rr-living-on-a-spike-food-210611-en_4.pdf>.

[12] www.unicef.org/nutrition/training/list.html.

IFPRI, 1995. Employment for poverty reduction and food security. von Braun, J. (Ed.), Washington, DC.

IFPRI, 2008. Food and financial crisis: implications for agriculture and the poor. Washington, DC. <www.ifpri.org/publication/food-and-financial-crises>.

Magen, B., Donovan, C., Kelly, V., 2009. Can cash transfers promote food security in the context of volatile commodity prices? Food, Agriculture and Natural Resource Policy Analysis Network. Executive Summary. Retrieved from: <http://www.fanrpan.org/documents/d00802/> (accessed 13.05.15.).

Maxwell, D.G., 1996. Measuring food insecurity: the frequency and severity of coping strategies. Food Policy. 21 (3), 291–303.

Micronutrient Initiative *et al.*, 2009. Investing in the future: a united call to action on vitamin and mineral deficiencies. Global Report. Micronutrient Initiative, Flour Fortification Initiative, USAID, World Bank, UNICEF, GAIN (Global Alliance for Improved Nutrition).

Tulchinsky, T.H., 2010. Micronutrient deficiency conditions: global health issues. Public Health Rev. 32, 243–255.

Shaw, J., 1995. Future directions for development and relief with food aid. In: von Braun, J. (Ed.), Employment for Poverty Reduction and Food Security. IFPRI, Washington, DC USA, pp. 252–274. (Chapter 10).

Chapter | Five

# Prevention of Future Food Insecurity

## 5.1 ANTICIPATING CRISES BY ASSESSING AND MITIGATING RISKS AND VULNERABILITY

### 5.1.1 Early Warning Systems

In mid-June 2015, the State News Agency of North Korea (Democratic People's Republic of Korea) issued an emergency warning of pending severe food shortages, because of the largely failed rains of 2014, the lowest over the previous 30 years. The News Agency reported that many of the State granaries were very low in stocks. Some 30% of rice paddies in the main rice-growing Provinces were parched, with the clay soil cracking. This official national alert follows the UN appeal of April 2015 for $US111 million emergency humanitarian aid, to a country which is dependent on foreign aid to feed its people, even when the short summer rains are favorable. About 70% of North Korea's 25 million people are food insecure and almost one-third of children under 5 years of age are stunted, the United Nations said in its report on North Korea "Humanitarian needs and priorities 2015", released before the funding appeal. Up to a million North Koreans are believed to have died during a widespread famine in the 1990s.

Food security early warning information systems have proven themselves useful and successful predictors of crises, in the Sahel and Horn of Africa, for example. FEWS NET, the Famine Early Warning Systems Network, is a leading provider of early warning and analysis on acute food insecurity. Created in 1985 by USAID, following devastating famines in East and West Africa, FEWS NET provides objective, evidence-based analysis to help government decision-makers and relief agencies plan for and respond to humanitarian crises. FEWS NET also supports and conducts training and capacity-building for national early warning systems, weather services and other agencies.

Analysts and specialists in 22 field offices work with US government science agencies, national government ministries, international agencies and NGOs to produce reports on more than 36 of the world's most food-insecure countries. FEWS NET has a vast network of partners, ranging from



Food Security in the Developing World. DOI: http://dx.doi.org/10.1016/B978-0-12-801594-0.00005-1

collaborators in data collection and analysis to readers of the reports. The latter include monthly reports and maps detailing current and projected food insecurity; timely alerts on emerging or likely crises; and, specialized reports on weather and climate, markets and trade, agricultural production, livelihoods, nutrition and food assistance.

For West Africa specifically, the regional early warning system of CILSS/ECOWAS/UEMOA[1] is a platform from which partner countries, donors, UN and civil society organizations pool information in order to conclude a joint analysis of regional food insecurity. It was instrumental in sounding the alarm in the early phases of the 2012 crisis there. This and other regional systems are continually being improved—the Integrated Phase Classification approach, for instance, allowing partner countries and regional institutions to prepare an appropriate response in advance of crises.

FAO operates an evidence-based global information system called the Food Insecurity and Vulnerability Information and Mapping Systems (FIVIMS).[2] Its purpose is to assist countries to characterize their food-insecure and vulnerable population groups, and use these data to advocate for national policies and programs to enhance food and nutrition security. FIVIMS also helps countries develop their Food Security Information and Early Warning System (FSIEWS), poverty maps, and food security and nutritional profiles.

There is still need for a more systematic link between such regional and national information systems to be linked with other databases on, for instance, child undernutrition, agricultural production, grain stocks, market dynamics and food prices.

### 5.1.2 Focusing on Prevention and Preparedness

The response of the international community and affected countries to mitigate the effects of drought and food insecurity crises has also brought into focus the need to both *prevent* such crises from happening and to build *preparedness* of individuals, households, countries and regions to properly manage the effects and consequences of such crises.

For this to come about, national and regional programs have to address and remove the underlying causes of *vulnerability, such as poverty, thereby instilling resilience*. Reduced vulnerability to shocks is only possible if it is embedded within partner countries' development policies, and for this to happen a "risk analysis" (including disaster risk management and food crisis management) needs to be integrated into national/regional policies. Additionally, institution building and capacity-building of personnel (of public sector and civil society especially) need to be prioritized, funded and implemented, geared

[1] *Comité permanent Inter-Etats de Lutte contre la Sécheresse au Sahel*/Economic Community of West African States/*Union Economique et Monétaire Ouest Africaine*.

[2] SICIAV in French and Spanish "*Systèmes d'information et de cartographie sur l'insécurité alimentaire et la vulnérabilité*"; "*Sistema de información y cartografía sobre la inseguridad alimentaria y la vulnerabilidad*".

toward disaster prevention and managing disasters if/when they occur. The potential for Public Private Partnerships needs to be explored.

Development Assistance these days is conditioned by two resolutions, of 2005 and 2008 (the Paris Declaration on Aid Effectiveness, and its subsequent review of progress), the prime mover for which was the OECD. Collectively, the EU is the world's largest provider of Official Development Assistance (ODA), amounting to EUR 56.5 billion in 2013, compared with the global amount from OECD/DAC donors of EUR 101.5 billion. The EU is also the most significant trading partner for developing countries, as well as a key source of technology, innovation, investment and entrepreneurship. For these reasons, Sections 5.1.2 and 5.1.3 concentrate on this particular donor.

The EU and its Member States are committed to increasing their action to meet the Millennium Development Goals (MDGs), particularly in order to eradicate extreme poverty and hunger in developing countries (MDG 1). Poverty reduction is the ultimate aim of EU development policy, as has been reaffirmed by the EU in the "Agenda for Change"[3], which prioritizes cooperation in sustainable agriculture, including the safeguarding of ecosystem services, and food and nutrition security. The EU has also enjoined the efforts of the international community to improve aid effectiveness, and has subscribed to several commitments including the Paris Declaration on Aid Effectiveness (2005), the Accra Agenda for Action (2008) and the Busan Partnership for Effective Development Cooperation (2011).

Because food insecurity is directly linked to poverty, both are priorities in the EU *Consensus on Development* of 2006.[4] In its responses to food and nutritional insufficiency, the EU Food Security Policy places food availability and access to food at the center of its poverty reduction efforts. The EU supports broad-based food security and poverty reduction strategies at both national and regional level, in preference to disbursing food aid directly.

EU actions address the structural causes of food insecurity and poverty at three levels—national, household and individual. In assessing and implementing support for food security initiatives, the EU complies with a number of imperatives, such as building and consolidating national capacity and networking; coherence with, and strengthening of, national policies and strategies for economic growth, poverty reduction and social cohesion; targeting the most marginalized in a community; improving incomes for the marginalized; and enabling sustainable outcomes of food security interventions whenever possible.

The *Agenda for Change* says that the EU should encourage more "inclusive and sustainable economic growth" which is "crucial to long-term poverty reduction". Building local institutions and business capacity, encouraging small and medium-sized enterprises (SMEs) and cooperatives, and creating regulatory and legislative framework reforms, can all promote growth.

---

[3] EC COM(2011) 637, as endorsed by May 14, 2012 Council Conclusions.

[4] http://ec.europa.eu/europeaid/what/development-policies/european-consensus/index_en.htm.

In China, for example, private sector growth has delivered millions of people from poverty.

All initiatives should be consistent with EU development policy and the Commission's country and regional support strategies, and there is close coordination with other EU financial instruments, ECHO, EU member countries and major donors, such as FAO[5], CGIAR, WFP and UNICEF, to ensure complementarity. The instruments of EU financial support of international food security include project support, budget support and (in emergencies) food aid.

*Project support* is provided in countries with a weak institutional framework where the policy environment does not allow budgetary aid. Support is designed to ensure *inter alia* that: (1) financial support is allocated to "food-insecure" groups; (2) development assistance is managed properly and capacity-building promoted where the public sector is weak; (3) key bottlenecks in food availability and access to food are mitigated; (4) aid recipients participate actively in project design and implementation; and (5) food security projects are supported for a limited time during transition from relief to long-term development, or when food insecurity is structural.

*Budget support* (being money paid directly to the recipient country's government to fund development projects) is provided under well-defined conditions, to low-income and least-developed countries, and where policies and programs exist to promote food security and reduce poverty. It is suitable only for countries with sound administrative capacity, and a reasonable record of pluralism and human rights. The aim of Budget Support is to (1) support policy and institutional reforms related to food security, (2) facilitate import of food by the private sector, (3) promote employment and income generation to improve access to food, and (4) help provide safety nets.

*Food aid* is considered a tool for humanitarian assistance, demand-driven and based on reliable, objective and transparent assessments of needs, while complying with broader food security strategies. In grant form, EU provides food aid only in acute protracted crises to meet well-identified and internationally recognized needs, where it is the best tool to solve an immediate problem. EU food aid is provided in consultation with ECHO and the main UN and NGO partners.

### 5.1.3 Decreasing Vulnerability and Building Resilience

Recent and recurrent food crises in the Sahel region and the Horn of Africa have underscored the need to address chronic *vulnerability* and develop a long-term and systematic approach to building the *resilience* of vulnerable countries and populations. Subsistence farmers, pastoralists and slum dwellers must feature strongly in the target groups. "Resilience" here is defined as "The ability of an individual, household, community, country or region to withstand, adapt and quickly recover from stresses and shocks".

[5] The 2004 EC-FAO Partnership.

In sub-Saharan Africa, the poorest households, communities and countries have suffered from a diminished capacity to recover from the long-term effects of climate change, especially frequent and intense droughts as well as economic crises and internal conflicts.

As discussed in chapter 3 "Causes of Food Insecurity", the effects of economic shocks, political instability and conflict, rising and fluctuating food prices, demographic pressure on resources, climate change, desertification and environmental degradation, insecure access to land and water, rural poverty, low productivity, weakened terms-of-trade, weak governance and insufficient investment in agriculture, have in many parts of the world (such as the Horn of Africa), resulted in greater exposure to risk and increased vulnerability. In the case of food insecurity, despite some progress around 800 million people are still suffering from hunger, the issue being particularly acute in drought-prone areas where most of the population depends directly on agriculture and pastoralism.

Resilience strategies of nations and their development partners, such as the EU, should contribute to a range of *policies*, in particular Food Security, Climate Change Adaptation[6] and Disaster Risk Reduction (DRR)[7].

The EU Food Security policy EC COM (2010) 127 presents *a new policy framework* to tackle hunger and undernutrition. The EC proposes that development strategies must address recent phenomena such as population growth, the food price crisis and the effects of climate change on agriculture. The EC intends to prioritize action in the *most fragile countries*, namely those which are most off-track in reaching the MDGs (in particular in Africa and South Asia). To increase action effectiveness, the Commission will support developing countries' national and regional policies, harmonize EU and Member States' interventions, and seek to improve the coherence of the international governance system

The concept of *resilience* has two dimensions—not merely the inherent strength of an individual, household, community or larger entity to better resist stress and shock, but the capacity of this entity to "bounce back" and recover relatively rapidly from the impact. *Enhancing resilience* lies at the interface of humanitarian and development assistance and calls for a *long-term approach and complementarity*, ensuring that short-term actions lay the groundwork for medium- and long-term interventions. The approach should be based on alleviating the underlying causes of crises *and* reducing intensity of the latter's impact through enhancing local/national/regional coping and adaptation mechanisms and capacities to better manage future shocks, uncertainty and change.

---

[6] EC COM (2009) 147 final White Paper "Adapting to Climate Change: towards a European framework for action".

[7] EU Strategy for Supporting Disaster Risk Reduction in Developing Countries EC COM (2009) 84 of February 23, 2009 "Towards an EU Response to Situations of Fragility: engaging in difficult environments for sustainable development, stability and peace" EC COM (2007) 643 of October 25, 2007.

As pointed out by the EC in its COM 586 document (see Section 4.6), resilience can only be constructed bottom-up. Partner countries, their policies and priorities, must take the lead. Action to strengthen resilience should be based on sound methodologies for risk and vulnerability assessments, the latter serving as the basis for elaborating and implementing national resilience strategies and disaster reduction management plans, as well as for designing specific projects and programs. A vulnerability analysis is needed before an intervention is planned let alone implemented, to optimize project design.

Clearly, with so many development partners active in food security and disaster relief operations, there is need for maximum sharing and complementarity of information and actions. For countries facing recurrent crises, the EU will work with host governments, other donors, regional and international organizations and other stakeholders to create platforms at country level for ensuring timely exchange of information and coordination of short-, medium- and long-term humanitarian and development actions to strengthen resilience. The EU will also promote resilience in international fora including the G8, G20, the Committee on World Food Security (CFS) and the Rio Conventions[8], and resilience will feature as a key theme in its partnerships with organizations such as FAO, IFAD and WFP.

In response to the massive recent food crises in Africa, two of the initiatives which the EC is pursuing are "Supporting Horn of African Resilience (SHARE)"[9] and "*l'Alliance Globale pour l'Initiative Résilience Sahel*" (AGIR)[10], which set out a new approach to strengthening the resilience of vulnerable populations through learning from experience, in order to multiply and scale-up successful approaches and actions. The SHARE and AGIR initiatives represent an improvement in the way humanitarian and development assistance interacts, boosting levels of assistance in the short-term, facilitating the link between relief, rehabilitation and development, and demonstrating the commitment of the EU to address the root causes of food insecurity in the longer term. They focus on food security in sub-Saharan Africa, but this approach can equally be applied to other regions and other types of vulnerability (eg, regions threatened by floods, cyclones, earthquakes, droughts, storm surges and tsunamis, climate change or food price increase).

During the 2011 drought, 3.7 million Kenyans were in immediate need of food, clean water and basic sanitation. To support Government of Kenya's efforts to strengthen institutions and increase investments in arid lands to better prepare the country to mitigate the impact of a similar future crisis, the EU mobilized resources under SHARE. Through this scheme, the EU enhanced its support to boost recovery and resilience building, in several

[8] United Nations Framework Convention on Climate Change, United Nations Convention on Biological Diversity and United Nations Convention to Combat Desertification.
[9] European Commission Staff Working Document SEC (2012) 102 of April 11, 2012.
[10] http://europa.eu/rapid/pressReleasesAction.do?reference=IP/12/613&format=HTML&aged=0&language=EN&.

ways—assistance to streamline the Kenyan early warning system; institutional support to the Ministry of Northern Kenya; support to the National Drought Management Authority (NDMA) in managing a Disaster and Drought Contingency Fund (NDDCF); enhancement of local authorities' capacities to manage an early response; and, community level livelihood projects expanding economic opportunities.

Investing in resilience is cost-effective. Addressing the root *causes* of recurrent crises is not only better than responding only to the *consequences* of crises, but is also much cheaper. All development partners are coming under increased pressure to deliver maximum impact for the funds available. In the cost-effectiveness context, the PSNP in Ethiopia, one of the largest social transfer schemes in sub-Saharan Africa, which provides for transfers in the form of food or cash to the most vulnerable households in the country in return for participation in public works, *costs only about one-third of an equivalent humanitarian intervention* (see Case Study 1 on the book's companion website, http://booksite.elsevier.com/9780128015940).

Yet such efforts at building resilience do not have to rely only on donor partners. A heart-warming community effort in Tigray province of Ethiopia is a case in point, as described in a BBC magazine article in April 2015 (BBC News, 2015). It was in Tigray that the terrible famine of the mid-1980s occurred. The tasks undertaken there over the last decade are akin to public works funded under PSNP, yet the terracing of hillsides cited in the article is solely using a "compulsory community labor" self-help initiative with no reward other than turning its degraded landscape into productive farmland. This community effort seems to have full ownership by the beneficiary communities, with tasks undertaken with joy and commitment.

## 5.2 NATIONAL FOOD SECURITY STRATEGIES

### 5.2.1 Background

Internationally funded "food security" programs need to reflect partner government and regional policies and processes, not only to address current challenges but also to anticipate future shocks and build resilience against them. To ensure this, food security strategies should be in place in each country, facilitated and endorsed by government. These should contain an analysis of the food security situation in the country, identification of the way in which threats to this could best be addressed, and an action and monitoring plan whereby this could happen, along with the actors involved, a timeline and budget.

Such strategies should be formulated in a participative way to assure national ownership, following a stakeholder analysis and full consultation with representatives of stakeholder groups identified therein (from the public and private sectors, civil society and donor community). The roles of each Working Group in achieving food security are complementary. Within *Civil Society*, NGOs are seen as a vital link to sensitize the citizenry about their rights and obligations, to train their members in the skills necessary to formulate their needs into a bankable document, and to assist in "project" implementation. Strategy implementation at "grass-roots level" is foreseen as being conducted by NGOs in close association with community-based organizations (CBOs). By contrast, the *Private Sector* is instrumental in providing services at both input and output levels, such as credit to enterprise development and marketing. Lastly, the *Public Sector*'s role is to provide an enabling legal and regulatory environment for development activities to flourish.

Similarly, with the action plan, before a strategic grain store[11] is planned let alone constructed, for instance, full consultation is needed with the community which constitutes the wider "user group" as to where it should best be sited, how it is to be managed and by whom, and what the maintenance and repair arrangements would be.

Consultation with communities (and others in-country) should reveal what they see as the primary challenge to their food security—perhaps upgrading a rural road, value-addition to the primary agricultural product, an effective commodity marketing operation put in place[12], local manufacture of portable mills to extract palm oil etc. These ideas can be fed into construction of investment/business proposals, and the most viable options taken to the Chamber of Commerce (for instance) to try to match business opportunities with potential investors. Agricultural processing ventures and smallholder machinery manufacture are often located in peri-urban areas, and the symbiotic links and potential synergies between rural and peri-urban areas need to be exploited.

Food security strategies should seek out and maximize the synergy between gender equality and food security. During the participatory development process, women's views need to be affirmatively sought, rather than predominantly men's views recorded, which would skew the formulation away from the perspectives of those who traditionally are responsible for home-making and decisions related to feeding the family and promoting their nutritional security. Such proactivity to secure the views of women is especially cogent in the public sector, where senior posts are usually held by men—young women take time out to raise children, and often return to their Ministries having sacrificed their places in the staff development hierarchy. Food security strategies should also support women's collectives, and adopt a rights-based approach in order to improve accountability and encourage independent monitoring of progress. In their action plans, food security strategies should be phased and multi-year, including components that can transform existing gender roles.

---

[11] As conceived by government, compared with a commercial grain store which would likely be planned by the private sector.

[12] As an example, in Liberia there is an effective value-adding and marketing structure in place for smallholder raw rubber, but not for cocoa beans. Cocoa production therefore remains at subsistence level whereas smallholder rubber is often "commercial". Smallholder palm oil marketing is at an intermediate stage of development.

### 5.2.2 Need for a National Strategy and Action Plan[13]

- Providing a framework for socioeconomic, technical and institutional causes of food insecurity to be progressively removed, rendering the citizenry less vulnerable (more resilient) to food insecurity, and remaining so.
- Providing a sustainable and coordinated "solution" to food insecurity in the country concerned, replacing a hitherto *ad hoc* approach[14], promoting synergies and avoiding wasteful duplication, and for components to be outcome-oriented rather than activity-oriented.
- Serving as the vehicle for implementing a Government's food security policy, where it has one, whereby citizens and other groups (these could include guest workers and refugees from other countries) in the country would be food-secure, and for Government to have a clear vision of *what* and *how* it intends to prioritize and coordinate implementation of that policy.
- Demonstrating that clear vision to donors and investors thereby assuring their commitment, so that strategy implementation is properly resourced and funds ring-fenced in advance.
- Setting the scene for related multi-sectoral planning and implementation at Governorate (or its equivalent in the country concerned) and Municipality level, and providing a mandate against which potential projects can be assessed as "bankable" and worthy of funding.
- Interventions to be identified by citizens/government and supported by donors, rather than the converse.
- Encouraging a development agenda with a preventative rather than curative orientation, together with a better-coordinated safety net of food security-related relief efforts in the event of disaster.
- Fostering *collective* and *focused* "action" over "rhetoric".

### 5.2.3 Strategy and Action Plan Outcomes

- Strengthened national capacity to effectively coordinate, manage and monitor implementation of food security measures.
- Sustainable institutional mechanisms installed and operative for policy decision-makers, in respect to technical support, communication and financing.
- Community, household and individual commitment expressed *through their own efforts*, to which value is added through buy-in and close collaboration, networking and complementarity among stakeholders—civil society (citizenry, NGOs, CBOs, unions), the public sector (central and

---

[13] A case can be made for regional food security strategies too.

[14] Some international agencies may already be implementing what they call a food security initiative, whilst other agencies may have ongoing projects that should also promote food security, even though that term may not be specifically used. The Strategy framework will enable all ongoing work related to 'food security', and that programed for the future, to be coordinated by the government concerned, in liaison with donors.

local governments), the commercial private sector and the international community (donors, UN bodies and other international organizations, development banks etc.).

- All categories of citizen receiving equitable consideration, especially the most disadvantaged, irrespective of political affiliation, gender or ethnicity.

### 5.2.4 Characteristics of the Strategy

#### 5.2.4.1 MANAGEMENT TOOL FOR GOVERNMENT

A national Food Security Strategy is a management tool for government decision-makers. They will arrange interventions that may be implemented in a multi-sectoral and multi-agency way, with responsibilities delegated as per agency mandates and obvious comparative advantages. The Strategy may also be viewed as a management tool for donors, who will identify in which category of intervention their own interests and comparative advantage lie.

Notionally, each entry in the Strategy charts[15] should be viewed as a "budget line", for funds need to be sought for each on the basis of "bankable project documents" in a competing environment, funds never being sufficient compared with overall need. Each project document should demonstrate its realism, its high probability of attainment and large positive impact on one or more aspects of food security, and how the project will be implemented in a given community, with maximum "ownership" and input in cash and/or kind by that community.

Simple and sensitive impact-indicators, and means of monitoring and evaluating any intervention (including food aid instruments), need to be identified for any "project" together with a convincing needs assessment to underpin each proposal, which would also have to demonstrate cost effectiveness.

The Strategy may be ambitious, but not overly so, for not all its intervention category provisions may be implemented, as some may not attract high enough prioritization. Yet the provisions of the Strategy need to be wide, against a time when the given country decides to do something not tried before, perhaps when the political context allows. Thus, in the Strategy charts, the term "marine enterprises" may be used rather than "fishery enterprises", a generic term that includes the growing of seaweed as a table salad rather than solely for fishing. Perhaps one day, seaweed culture may be a viable and profitable enterprise for a country with a coastline, just as it is in the Philippines and has been considered as an option off the coast of Saudi Arabia, and the National Food Security Strategy concerned will "enable" it to qualify as a candidate for development funds.

---

[15] There could be four charts, one each for measures which promote "availability", "access", and "use" and one for the enabling institutional setup.

#### 5.2.4.2 COMPATIBILITY WITH OTHER GOVERNMENT POLICIES AND STRATEGIES

The Strategy should be compatible and convergent with all major plans, policies, strategies and any other multisectoral initiative of Ministries, and be generally acceptable to civil society and private sector umbrella organizations.

#### 5.2.4.3 DEVELOPMENT AGENDA

The Strategy should encourage removal of both the socioeconomic and technical primary determinants of food insecurity, while enabling exploitation of opportunities to become more food secure. This composite goal will best be achieved by promoting a development agenda as a continuum to an agenda which may currently address mainly relief and rehabilitation.

#### 5.2.4.4 FLEXIBILITY

The Strategy should be flexible, capable of modification in the light of experience, and able to absorb and adapt to lessons learnt.

#### 5.2.4.5 LIVELIHOODS AND TARGETING

The Strategy should have roots in the current livelihoods of beneficiaries. It needs to target in particular the most geographically, socially and economically disadvantaged areas/households/individuals, perhaps living at the margin of the market economy. The latter should be adequately informed of the provisions of the Strategy, such that they demand their needs be met through it, and indeed that mobilized public opinion becomes the driving force behind its implementation.

### 5.2.5 Institutional Arrangements for Implementation

National food security governance must remain the overall responsibility of the government, working closely with civil society and the private sector to implement the Strategy through complementary task-oriented action plans. The seminal issue is the identification of a national body responsible for the coordination, supervision, planning, monitoring and evaluation of that implementation. The Strategy must be nested in an institutional structure within government that will be effective in coordinating its implementation, perhaps to be called the Food Security Steering Committee or the National Food Security Commission. Implementation will be long-term, rather than a short-term "fix". Effective implementation is conditional on the Steering Committee being constituted at the appropriate high level of representation, mandated and with powers of coercion, and properly informed and resourced. If possible, the institutional structure should not involve the creation of a "new" body but adapt to an existing one which is currently "performing". The ideal would be that the Steering Committee proves so useful and therefore "in demand" that it becomes self-financing from national resources (viz., independent of donor support).[16]

The institutional structure should enable the government to coordinate activities both in the Center and at Governorate and Municipality level, and engage in reciprocal consultation with donors and UN/international organizations. The government needs to skilfully handle the moderating and coordination oversight role within the Steering Committee. It needs to merge the perspectives of different line Ministries without impinging on their need to oversee projects within their sectors, and foster inter-Ministry/Agency cooperation on any project funded under the Food Security Strategy umbrella.

To vitalize donor coordination by the government lead agency, a multisectoral Food Security Sector Working Group (FSSWG) is envisaged, which would gather key line ministries, UN agencies, donors, representatives from International NGOs and civil society, to offer technical support to the Steering Committee.

*For further information on a participatory Strategy, the reader is referred to the Palestine National Food Security Strategy of 2005 (*Ashley and Jayousi, 2006*), and related Action Plan of 2006, which became ensnared in political inaction following the election of 2006. The Strategy contains charts of Strategic objectives and detailed means whereby these may be achieved. An example of a well-thought through national food security strategy in Africa is that of South Africa (2002)* (www.gov.za/documents/integrated-food-security-strategy-south-africa).

## 5.3 ADDRESSING FOOD AVAILABILITY

### 5.3.1 Through Exploiting Natural Resources

#### 5.3.1.1 INTRODUCTION

During their growth, both crops and livestock require good management, including proper nutrition and protection against pests and diseases, if productivity (per unit of land, water, labor or investment) and profitability are to be optimized. Pest and disease practices are best applied to "manage" rather than "control", this leading to minimal compromise of food safety and environmental quality; laboratories are crucial for adequate checks on this being made, so conforming to best practices.

Land use should be rational, with arable crops on the best land, in general, and forestry enterprises on land which cannot support profitable arable farming. The world's drylands are often best suited to extensive livestock production, though stocking rates are often higher than optimal, a result of demographic, conflict and/or climate change factors. Where there is adequate ground or surface water, irrigation can bring about massive yield increases in field and fodder

[16] Using performance-based satisfactory outcome-dependent cost-recovery modalities to cover fixed and operational costs, perhaps with contributions from the beneficiary communities/private sector.

crops. Land quality and water-holding capacity can be improved with suitable interventions, and cultivation techniques which minimize soil loss through erosion and evaporation of water from tilled surfaces. Conservation agriculture (see Case Study 2 on the book's companion website), as is becoming popular in southern Africa, for instance, promotes this through the use of minimum tillage equipment, while reducing weed infestations and labor requirements, mainly of women who do much of the agricultural manual work. Adaptive strategies by farmers to cope with their soil and climatic conditions are widespread, as for instance in the dry northeast corner of Nigeria (Ashley and Shugaba, 1994), and reference is made to them also in the case studies for Belize (see Case Study 3) and Lao (see Case Study 6).

For subsistence farming to transform into semi-commercial farming and beyond, active farmer associations can be instrumental, with the added value of economies of scope and scale in purchasing inputs and produce marketing. The commercial private sector is crucial for this too, with safeguards against malpractices like adulteration of agrochemicals or rigging of prices. To safeguard against the latter is often the role of the public sector, devising laws, regulations and standards, and enforcing them. Unless agriculture is market-led, with effective storage areas, roads and transport, it will not be profitable. Afforestation leads to erosion control, stabilization of the water cycle and, like agricultural and livestock primary and secondary production, wealth creation which fosters the accessibility aspect of food security. Natural forest ecosystems provide an enormous amount of food (as well as firewood, medicines and income) to complement that from traditional agriculture, thereby fostering food and nutritional security. This provision is detailed in a report from 60 leading scientists which calls for a global commitment to protect forests (Collaborative Partnership on Forests, 2014).

Six examples of adding value to the food production potential of land and water are given in the following sections.

#### 5.3.1.2 THE GREEN REVOLUTION

Seed security is essential to ensure food security. Ever since humankind started adopting a more settled form of crop agriculture, farmers have selected seed of individual edible plants to grow on particular patches of "farmland", initially sourced from the wild. Within these stands of crops, what seemed to be the "best" individuals were selected at harvest time to provide planting material for the following season. Such selection of seed and careful storage of it (often in the rafters of the thatched homestead where the smoke from the hearth controls insect pests) has provided the basis for crop improvement. Similar practices occurred with livestock, with wild animals being captured and raised in settlements or herded extensively, with the best individuals used as breeding stock.

Over time this advancement in both crop productivity and production was lent weight by agricultural science. This was harnessed to improve not only the genetic and physiological base of crops to confer a range of high-yielding disease-resistant characteristics, but also the agronomic base. The latter has better enabled plants to access the minerals and water they have need of to grow well and be protected from competition for these by weeds, while disease and pest organisms have been tackled through natural resistance, agrochemicals and integrated pest management (IPM). Similarly with livestock, there have been initiatives to improve the genetic and physiological base, and better management to enable elite genetic traits to be most fully expressed in growth rate, offtake weight and fertility.

These modern efforts have been rooted in national (government, university, community and private sector) and international agricultural initiatives. A quantum leap in crop productivity was observed in developing countries with the Green Revolution (GR) (so-called by the then-Director of USAID). Though this work started soon after the Second World War, the term was first used in 1968 referring to an initiative which started in the late 1940s, led by an American crop scientist Norman Borlaug who was working with an international group called CIMMYT (International Maize and Wheat Improvement Center) in Mexico. GR in developing countries has received strong support over the years by the Ford and Rockefeller Foundations, World Bank, CGIAR, USAID, USDA, FAO, IFAD and UNDP in particular. The "revolution" comprises a series of agricultural research and development initiatives in developed and developing countries. This involved the creation of potentially high-yielding cereal cultivars which respond to fertilizers producing a higher Harvest Index (ratio of the commercial unit to total plant biomass), and having day length (photoperiod)-insensitivity, broadening the scope of their success at different latitudes. Employing high-input agronomic practices was the second component of the success "package". Synthetic inorganic fertilizers, and pesticides, fungicides and herbicides were mainstreamed in agriculture for the first time in the middle 20th century, and played a major role in the GR, as did agricultural mechanization. Combined, the genetic and agronomic components enabled crop productivity per unit area, per season and per year (using multiple cropping), and per unit of labor, to be greatly enhanced, and hence these new varieties were deemed valuable and "improved" by participating farmers, compared with their "control" cultivars/landraces, and adopted by them.

The GR started in Mexico, spreading to southern Asia (particularly India, Pakistan, and Philippines) in the 1960s and 1970s, and from there to China over the next two decades. It was the famine and conflict on the Indian subcontinent of the mid-1960s which created the local demand for the Borlaug team's intervention in Asia, following its initial success in South America. Africa was relatively unaffected by the GR of the 1960s–70s, in part because of resistance to high-input high-yielding varieties by international environmental groups, and also to constraints in infrastructure and human capacity, and corruption. In the early 1980s, this resulted in the World Bank and Rockefeller and Ford Foundations withhold funding of African projects in which Dr Borlaug was involved. However, by 1984 a significant start was made in Ethiopia prompted by the ongoing famine there (and low incidence of corruption), which led to uptake of genetic and agronomic advances, and a slew of other African countries followed, initially in the more developed economies, with the help of the Carter Center and the Sasakawa Africa Foundation.

The genetic initiatives promoted during the early GR involved cultivars and hybrids of cereals (mainly wheat, rice and corn) identified from within the international community, further adapted to local conditions in the country concerned, through crossing with local types having yield-enhancing characteristics. Simultaneously, agronomic practices were enhanced in the beneficiary country, through providing irrigation water, mineral fertilizers and other agrochemicals. All these inputs led to massive productivity increases per unit of existing agricultural land, and opened the way for additional land to be brought into production. In Brazil for instance, the extensive liming of acidic unproductive soils enabled a soybean and cattle agro-industry to flourish. Brazil is now a significant exporter of soya, beef and poultry.

In Mexico, Borlaug's team bred high-yielding disease-resistant semi-dwarf wheat cultivars. Mexican farmers had been suffering huge yield and profitability losses up until then, particularly due to rust disease. It was necessary to incorporate a dwarfing component in the genotype to counter the vegetative response to high nitrogen application to the soil. With the heads of the new wheat cultivars being so much heavier, a tall plant would have fallen over (lodged). The rewarding experience in Mexico led to expansion of the program to India, Pakistan and Turkey, where again introduction of dwarf spring wheat cultivars was successful. The work with wheat also led the way to the development of high-yielding semi-dwarf rice cultivars at IRRI in the Philippines, such as IR8, this cultivar becoming popular in India and later adopted across Asia.

There is no date when the GR "ended", and it may be considered to be continuing, yet with a greater awareness of applying lessons from earlier decades to prevent or mitigate some negative effects that in retrospect have been acknowledged. Initiatives, such as conservation farming and/or agricultural input subsidies (in Zambia, Zimbabwe, Malawi, South Africa and Tanzania, for instance) and the introduction of NERICA rice (in Guinea) are examples of ongoing success stories.

"Genetic engineering" through transferring a trait into the DNA of a crop species that it does not normally possess has been decried by many in the environmental lobby, though it does hold tremendous promise for improving nutritional quality of grain, for instance. Developing countries which in 2013 had GM crops, which can be used for food (mainly corn and soybean) over more than 1 million hectares, were Brazil, Argentina, China, Paraguay, South Africa, Uruguay and Bolivia (ISAAA, 2013).

*Yield increases:* According to FAOSTAT, *wheat* yields in all developing countries rose during 1950–2004 from 750 to 2,750 kg/ha, on average.

Disaggregating the data, *Mexico* shows the most startling improvement (from 750 to 4,750 kg/ha) and *India* and *Pakistan* from about 250 to 2,500 kg/ha). At the start of the CIMMYT initiative in the 1950s, Mexico was a massive importer of wheat, though by 1963 had become a net exporter of that grain. From 1965 to 1970, wheat yields nearly doubled in both India and Pakistan—in India the national production of wheat in 1965 was 12.3 million tonnes(mt), rising to 20.1 mt by 1970, and by the year 2000 it was 76.4 mt, largely through greatly increased productivity per unit of land. By 1974 India was self-sufficient in the production of all cereals. In Pakistan, wheat yield was 4.6 mt in 1965, rising to 7.3 mt in 1970 and 21 mt by the year 2000, well beyond self-sufficiency.

In 1994 former US President Carter obtained Ethiopian President Zenawi'a support for a campaign seeking to aid farmers, using the fertilizer Diammonium Phosphate and Borlaug's methods, and in the following season, *Ethiopia* achieved the highest yields of major grain crops in the country's history, with participating farmers achieving a 15% average yield increase over the previous season.

As for *rice*, in *India* yields were around 2 t/ha in the 1960s, rising to 6 t/ha by the mid-1990s, which drove down the cost of the commodity on the local market several-fold, thereby improving economic access by the majority poor. Cultivar IR8 was hugely successful in India, and also in the *Philippines* where yield rose from 3.7 to 7.7 mt nationally over two decades.

*The outputs and impact of the Green Revolution:* Borlaug's GR has divided public opinion, and despite the clear benefits has its detractors, based on "unsustainability" paradigms. It has to be said, however, that many a detractor has never experienced hunger or poverty, let alone died of it or seen his/her children do so.

### 5.3.1.2.1 The Upside

1. Through increased productivity per unit of land, the poorest communities in many developing countries were able to eat more, a result of improved food availability, which also lowered prices to more affordable levels. Borlaug himself estimated that as a consequence of GR over a 40-year period, the proportion of hungry people in the world declined from about 60% in 1960 to 17% in 2000 (Borlaug, 2007).
2. GR provided grain which the rapidly increasing populations needed, without which there would certainly have been a greater incidence of hunger and suffering than did occur, associated with the droughts of the 1980s and subsequent years. It is often said that Borlaug saved the lives of a billion people.
3. Greater productivity will likely have saved much natural forest from being cut to bring incremental land into production to cater for the burgeoning population in the developing world. If the global cereal yields of 1950 had still prevailed in the year 2000, Borlaug estimated that an additional 1.2 billion hectares of land of the same quality would have been needed

rather than the 660 million hectares actually used to achieve the global harvest in 2000. Economists have dubbed this concept the Borlaug Hypothesis, namely, that "increasing the productivity of agriculture on the best farmland can help control deforestation by reducing the demand for new farmland".

##### 5.3.1.2.2 The Downside

1. Pollution of land and waterways with agrochemicals, causing (a) morbidity and death among rural communities, as documented in the Punjab, India, for example, (b) reduced numbers of fauna which may be regarded as "farmers' friends" (being predators or parasites of crop and livestock pests), (c) by destroying the "life" of the soil and removing "weeds" erosion of topsoil was encouraged, and (d) algal blooms from nitrogen fertilizer runoff, such as noted by Beman et al. (2005) in Mexico's Sea of Cortez. Those latter authors forecast that by the year 2050, 27–59% of all nitrogen fertilizer used will be applied in developing regions located upstream of nitrogen-deficient marine ecosystems, and they highlight the present and future vulnerability of these ecosystems to agricultural runoff.
2. Reduced biodiversity and genetic buffering against pest/disease/drought within the crop concerned, as local landraces were replaced by relatively few "improved" cultivars. In India, for example, there were some 30,000 rice cultivars/land races prior to the Green Revolution, whereas today there are only 10 or so major cultivars. The increased crop homogeneity has reduced the buffer of the crop to a disease outbreak, despite the multiline breeding techniques employed intended to confer disease resistance. This has led to the increased need for, and use of, fungicides.
3. The GR early approach has been criticized for bringing large-scale monoculture high-input agricultural techniques to countries that had previously relied on polyculture "intercropped" subsistence methods, which are more effective in utilizing soil nutrients and moisture when they are scarce.
4. Little emphasis in the early GR was given to grain legumes (pulses), which provide much of the protein in rural smallholder communities; thus hunger and food security was targeted more than was nutritional security. Much of the land used for grain legumes that fed Indian subsistence smallholders was replaced by wheat which did not make up a large portion of their diet. A lot of this wheat was consumed by urban communities, or exported to the world market for grain or animal feed. Food and nutrition security for smallholders was therefore compromised. The reduced availability of pulses from subsistence agriculture has been blamed for the increase in undernutrition, the other side of the GR reduced-hunger coin. (In recent years there has been increased emphasis on grain legumes, in programs at the International Center for Tropical Agriculture (CIAT) and the International Center for Agricultural Research in the Dry Areas (ICARDA) for instance).

*It should be said that early GR improvements to the genetic base in wheat and rice was far easier than for grain legumes and corn. For wheat and rice, scientists started with a rough "blueprint" for what was required, and had a large stock of improved germplasm from temperate zones, such as variety Norin 2 wheat, a cultivar descended from Japanese semi-dwarfs. However, there was little improved germplasm available for beans or lentils (or barley, sorghum, millet, potato or cassava). Even for corn, the more complicated mechanics of breeding and the high location-specificity of varietal technology made it difficult to adapt improved lines from the temperate zones to tropical climates and disease environments* (Evenson, 2003). *So, that grain legumes were not at the forefront of GR advances cannot be blamed on Borlaug and his team.*

5. As some micronutrient components of fertilizers are in relative short supply in the world, such as natural phosphatic rock, the increased use of this hastens the time when there will be none left or it is uneconomic to mine. Nitrogen fertilizers are products of the hydrocarbon industry, and oil and gas too are finite resources, their mining being unsustainable and their use responsible for greenhouse gas emissions and global warming.
6. GR agronomic techniques incur a high energy input, relying as they do on fertilizers, pesticides, herbicides and mechanization (rather than human labor and animal draft), and therefore on oil and gas extraction. A fall in global oil and gas supplies (therefore higher price) would constrain high-input agriculture, and could cause further food price spikes in coming decades. (*This negative has gained prominence in recent times, yet was not foreseen even by environmentalists 50 years ago*).
7. The liming of the extensive *cerrado* region in the Brazilian interior to render the soil fit for agriculture led to the destruction of great swathes of Amazonian forest, with consequent negative sociological and environmental fallout (see Case Study 5 on the book's companion website). (*This negative arose from the way the liming regime expanded "out of control" under the influence of multinationals and their lobbyists*).
8. Socioeconomic criticisms: (a) lowered food prices through greater food availability reduce farming profitability and can put the less efficient and smaller farmers out of business; (b) the transition from traditional agriculture, in which inputs were generated on-farm, to Green Revolution agriculture, which required the purchase of inputs, led to widespread establishment of rural credit institutions. Because wealthier farmers had better access to credit and land, GR increased class disparities. Smaller farmers often went into debt, resulting in forfeit of their farmland; (c) the increased level of mechanization on larger farms under GR reduced the need for manual labor, and created impoverishment of non-landowners, itself resulting in urban drift, swelling the numbers in urban slums; and (d) farming techniques that increase productivity are seen as benefiting large agribusiness corporations, and have also been cited for widening social inequality in the countries which had hitherto strived to bring about

more social equity through introducing land reform (see Section 6.3), such as Pakistan under the Zulfikar Ali Bhutto regime in 1972. Critics of GR say that its technological imperatives weakened socialist movements in many nations, and that this was the main rationale of the GR initiative, a devious ploy of free market geopolitics—in India, Mexico and the Philippines, for example.

*Resolution of the debate:* Some of the above criticisms may be discounted as unfair, and not a justification for the GR approach of the 1950s never having been attempted. Let the reader decide. Additional criticisms are definitely unfair. For instance, that the increased rural infrastructure needed to handle transport and haulage of the larger harvests were bad for the environment ! And that a lot of the high-yielding corn goes to produce biofuel *now*, is not a valid criticism that can be laid at the door of the early GR work. That productivity gains have not been shared equitably and that close to a billion people alive today have not significantly experienced its benefits are true enough, but that cannot be a criticism of the Borlaug method.

As Evenson (2003) *ibid.* says, the political economy of food security has two dimensions, global and local. From a global perspective, GR has been an extraordinary success. Food production has increased over a period of unprecedented population growth, and a greater proportion of the world's people are better-fed. By contrast, from the local perspective, huge numbers of people still do not have adequate diets while having low incomes, under $2 a day. After five decades of GR and international development programing, many developing countries remain in mass poverty. However, at the end of his paper, Evenson says "for the poorest countries, the Green Revolution, late and uneven as it was, was 'the only game in town'".

Borlaug accepted that the 1950s–80s GR had been but a start in the right direction, has not produced a Utopian outcome and that there is some way to go to balance massive ongoing human population growth, sufficient nourishing food with which to feed them and sustainable production methods—a work in progress. For instance, future world food security could have been under threat had some of the initial Green Revolution practices not been modified over recent decades to be more environmentally friendly (again, a work in progress). Borlaug appreciated some of the concerns of the environmental lobby, while pointing out that he regarded many of its advocates "elitist", who have never themselves experienced hunger and could afford the high prices that "organic" commodities need to attract to sustain low-productivity "organic" farming livelihoods.

#### 5.3.1.3 PARTICIPATORY VARIETY SELECTION

In developing countries, with resource-poor farmers inadequately served by public services, crop variety grain yield trials laid out on research stations, under researcher-managed conditions, are often found to give results which are of only limited use to the target farmers.[17] That may be for one or more reasons:

- The trial is conducted under high-fertility conditions, and other elite management practices (good disease and pest control, supplementary irrigation, timely weeding, etc.) which bear little resemblance to conditions on subsistence farmer fields under farmers' management in the target area. Thus a variety that performs "the best" in such a trial may not be "the best" on marginal farmers' fields.
- Such an on-station trial will likely test "elite" (in the researchers' opinions) varieties[18] against each other, without also testing in the same trial the "control" "variety" which a given farmer uses, so the farmer would not know whether the "best" entry yielded more than the one she/he normally uses.
- It is often not just grain yield which is important to farmers/households—other factors may be as much, even more, valued than yield. These include *field characteristics* such as stability of yield over a number of seasons under farmers' conditions; degree of drought tolerance; degree of resistance to diseases/pests/lodging; maturity period (which needs to match that of the rainy season); the amount of "straw" which a variety produces (which may be used for livestock fodder, thatching or fence building). *Postharvest characteristics* are also hugely important to farmers—does the grain have the needed high tolerance to pests in the granary, how long the grain takes to cook (and hence amount of firewood that needs to be collected by women and girls) and when prepared as food does the grain combat hunger (does it make one feel "full"?). All these field and postharvest characteristics need to be assessed, over a number of sites and seasons, under farmer-managed conditions, before a variety can be considered as good as, or better than, the one the farmer is currently growing.
- For these field and postharvest characteristics to be properly assessed, it is the farmer who is the best judge not a researcher(s) working on a research station, sometimes with very little interest in going to the field, or the means/budget to do so. Preceding any trial the farmers must first be consulted—"Which crop of those you grow do you most want improving, and which characteristics of your current 'variety' of a given crop do you want changed or retained?". Indeed, the whole household needs to make such decisions. Owing to different perspectives, what the male farmer may wish for may well be different from what the female farmer (wife) considers important (see Section 6.2.2). So, both males and females

[17] The example of grain crops is given here. Similar scenarios apply to nongrain crops, fodder crops, vegetables, small livestock, etc. More information can be located on the PVS technique in research papers by J. R. Witcombe and D. Harris, for example.

[18] For the purpose of simplicity, the term "variety" will be used throughout this description of PVS, even though the technical terms "landrace" or "entry" may sometimes be more appropriate.

need to be consulted as a pre-requisite of variety testing, and to assess the results. A good sample of farmers needs to be asked and the ideal "*ideotype*" (holistic set of characteristics) of a given crop then constructed, which can then be sought in a "project" situation (Donald, 1968).

#### 5.3.1.3.1 The Protocol and Benefits of Participatory Variety Selection

The participatory variety selection (PVS) protocol has proven itself in many countries. The success of PVS largely derives from its participatory nature—farmers, communities and research centers. It provides for a farmer-centered demand-based selection of what is truly valued by the farming community. An intervention can be made through "a project", whereby a small team of specialists in various aspects of agricultural science can facilitate beneficial changes in productivity and production. This can lead to households becoming self-sufficient in basic foodstuffs, and thereafter lifting a community out of poverty by generating a surplus of produce for sale.

The "project" is there as part of the start-up process; once the PVS system is properly under way, project staff can move on to other farmers elsewhere. The crop improvement process it leaves behind is self-sustaining, and can be upscaled by farmer-to-farmer referral within the area.

#### 5.3.1.3.2 The Step-Wise Way in Which PVS Works, in Summary

1. Following initial discussions with village leaders/opinion makers in the target area to introduce the concept, participating communities are selected in a (socially re-enforcing) "cluster", based on the degree of enthusiasm generated and willingness to actively participate in the project.
2. Election within those communities of project contact persons ("Village Extension Workers"—VEWs)[19], based on a set of criteria such as agricultural credibility, reliability and trustworthiness.
3. Selection of priority crop(s) and "*ideotypes*" (based on the expressed needful modifications of currently used 'varieties').
4. Project staff identify from local/national/regional/international research bodies[20] (and through field inspection visits to the various agroecological

---

[19] These VEWS can be paid a small fee by the project initially, until their services become recognised as important by their client farmers, who can then pay the VEWs in cash or kind, as there is an opportunity cost to the VEWS, their having less time to manage their own farms.

[20] The most appropriate international research center for many drought-prone areas in Africa and south and east Asia would likely be ICARDA, which specialises in crops such as chickpea, lentils, *Faba* bean, barley, wheat and fodder plants. This Center is based in Beirut, Lebanon. The Center can send a package containing seeds of a number of entries of a given crop, selected for their likely adaptation to the dryland country concerned. There would be an obligation for "a project" to report to ICARDA the comparative performance of the entries tested.

areas in the country concerned), "varieties" or landraces adapted to the farmers' conditions and requirements (Ashley and Khatiwada, 1992).[21]

5. Procuring sufficient seed of these varieties to distribute small samples "free" to participating communities for them to test under their farm conditions, under supervision by the project.
6. Training of VEWs in laying out simple on-farm trials on fields in a scientific way.
7. Procuring seed of "control" "variety/ies".
8. Making plan for testing say five varieties against the control per given crop on a small area of each participating farmer's land.
9. Packaging seed in bags, sufficient to plant plots of say $3\ m \times 2\ m$[22] in a "randomized block" design.
10. Distribution of these seeds to farmers by the VEWs who will supervise them planting the trial (at the same time as the farmer plants her/his farm with the usual "control" seed).
11. VEWs (with the support of project staff) will monitor the trials throughout the growing season, record against a checklist the key characteristics of individual varieties being tested, and collect farmers' opinions on them during the growing season and at harvest.
12. VEWs and farmers will dry, weigh and record individual plot yields, so that comparative yields can be compared per farm and across the whole area. Ditto postharvest characteristics of the preferred varieties.
13. From the data collected, a picture will emerge of variety preferences across the wider area.
14. Seed of the preferred variety(ies) can be multiplied/collected and tested on larger areas in the subsequent season(s) to confirm or otherwise its/their yield potential and other qualitative/quantitative characteristics.
15. Seed of the preferred variety/varieties can be procured/multiplied under contract and distributed at cost throughout the area in response to farmer demand.[23]

*Case example:* Under an EU-funded project 1996–98, the current author's project achieved success with applying this technique under farmers' conditions in Argakhanchi district in the low hills of western Nepal with both corn and oil

[21] This paper records an instance in which the district officer of the Ministry of Agriculture in Bhojpur District, eastern Nepal, assured the authors that there was nothing of agricultural interest to be seen in his district. However, within a short distance was found a field of wheat yielding 2 t/ha, three times as much as the average yield on rainfed land in the area. Seed from this homogeneous cultivar was procured from the farmer in barter exchange for an equal quantity of milled rice carried by porters from the nearest trading center (as the area was too remote to be part of the moneyed economy), and used as a test entry in PVS trials in western Nepal.

[22] The total area of the trial should take only a small proportion of the farmer's land, representing a low risk to his/her livelihood, and the potential of significant gain in knowledge.

[23] The purchase of an "improved variety" is the key indicator of its suitability for a given farmer in a given location.

seed rape. A corn composite called Hetauda Composite, sourced from the government research station at Hetauda, was selected by farmers as a type they preferred to their "local", on the basis of simple multilocational 10-entry trials under farmer conditions. In the case of oilseed rape, "Pakhribas local" sourced from near Pakhribas Agricultural Research Center in the low hills of Dankuta district of eastern region of Nepal proved superior to the farmers' "control" in Arghakhanchi district, being more resistant to lodging and disease, growing more robustly, yielding more highly and the oil having a "better" taste.

#### 5.3.1.4 PHYSIOLOGICAL AND BIOCHEMICAL RESEARCH

As well as the genetic and agronomic ways to improve crop yields, as discussed earlier, crop physiology can provide a useful tool as well, to guide breeders toward exploiting beneficial features inherent in a genotype, giving better tolerance to water and heat stress, for instance (Ashley, 1999). Many examples could be given of work under way, some having an obvious application to yield or nutritional improvement (see Section 5.6), and others which may appear at first glance to be more in the realm of "pure research". One of the latter is a research program regarding a colored spot pattern on flowers of daisies. Allan Ellis at Stellenbosch University, South Africa, and Steve Johnson at the University of KwaZulu-Natal in Pietermaritzburg have shown that petals of the Beetle Daisy *Gorteria diffusa* strongly resemble the bodies of female bombyliid flies and cause male flies to try to copulate with the daisy flowers. The petals have a complicated petal spot color (due to anthocyanin pigmentation) and associated epidermal elaboration that mimics the appearance of the plant's pollinating insect, something seen before only in orchids. Doctoral student Greg Mellers, working under Professor Beverley Glover, Director of the Cambridge University Botanic Garden, is trying to understand more about how this evolutionary mechanism has happened, at the genetic and molecular level. Such knowledge could perhaps then be put to use in some insect-pollinated crop plants, to increase pollen transfer, fruit set and crop yield[24].

#### 5.3.1.5 BIOLOGICAL CONTROL OF CROP AND FOREST PESTS, AND HUMAN/LIVESTOCK DISEASE VECTORS

One example of developmental research work on IPM is given here. Professor Tariq Butt leads a team of scientists at Swansea University in Wales, developing environmentally friendly strategies and products for arthropod (insects, ticks and mites) pest control. The Biocontrol and Natural Products Group (BANP) has developed entomopathogenic fungi (EPF) to control insect pests of socioeconomic importance, including crop pests (eg, weevils, aphids and thrips), vectors of human and livestock diseases

---

[24] Cambridge University Strategic Research Initiative in Global Food Security aims to integrate research ongoing across the various university Schools and Departments, to improve understanding of the challenges of food security, and to develop solutions (www.globalfood.cam.ac.uk).

**FIGURE 5.1** An adult specimen of the Large Pine Weevil *Hylobius abietis*, infected with *Metarhizium anisopliae* fungus. Fluffy white fungal growth (mycelium) can be seen exuding onto the surface from infected internal organs. This pest, which is present across Eurasia, can wipe out conifer plantations and reforestation efforts. Conifers provide wood for cooking (and heating) for poor people in the northern part of North Korea (DPRK), for example, and hence directly enable human food security. Photo: Tariq M Butt.

(eg, mosquitoes, midges and ticks), and a pest of forest trees (see Fig. 5.1). All these pests, vectors, and diseases negatively affect human food security.

The products based on these fungi are highly specific, targeting particular arthropods, which cause damage through their own activities or acting as vectors of pathogenic diseases. The EPF work in concert with predators and parasitoids in supressing pest populations. The products containing the pathogenic fungus are useful where no control has ever been attempted or where pests are resistant to chemical pesticides, and serve as a replacement for chemicals which are being phased out in line with new EU legislation and directives, notably EC 1107/2009 regulation and Directive 2009/128/EC which obliges EU Member States to implement principles of IPM, with priority to be given to non-chemical methods of pest control.

Most European food supermarket chains comply with such imperatives when sourcing foods from developing countries, endeavoring to provide healthier food with minimal or no chemical pesticides used in their production. Growers in developing countries are thereby rendered more competitive, spending less on broad-spectrum insecticides, while attracting premium prices (which bode well for the access component of their food security).

In addition to the EPF strategy, BANP follows a "push-pull" pest control strategy by developing *repellents* to prevent pest infestations and *attractants* for mass trapping and for deployment in "*lure and kill*" pest control programmes. A third strategy followed is that of "stress and kill", using low doses of standard insecticides or botanical deterrents to reduce the vigor of the pest, and increase its susceptibility to EPF.

Crop commodity focus has been on oilseed rape, cereals, potatoes and pasture, together with soft fruit and protected environment high-value vegetables and fruit. In the case of livestock, disease vectors targeted have included ticks and midges. Tick-borne pathogens affect 80% of the world's cattle population and are widely distributed throughout the world, particularly in the tropics and subtropics. Tick-borne diseases rank highly in terms of their impact on the livelihood of resource-poor farming communities in developing countries, particularly in parts of sub-Saharan Africa, Asia and Latin America where the demand for livestock products is increasing rapidly.

Another target for the BANP group is biting midges. One associated disease is bluetongue, an acute virus disease of ruminants, especially sheep, with high morbidity and mortality consequences. The disease is significant in South Africa and present in most other countries on the continent, in Asia, the Middle East, etc. The vector is almost exclusively the female adult hematophagous biting midge belonging to the *Culicoides* genus.

BANP has identified *Metarhizium* strains which kill both adult and larvae of this midge, marketed in the United Kingdom as *Metarhizium* MET 52 (granules), and other strains which are highly pathogenic to midges, and several tick and weevil species. Professor Butt has worked with industry and regulators in resolving commercial (product development), regulatory and release issues for the Group's biopesticides, and with research institutions in developing countries, such as ATBU, Bauchi State Nigeria.

#### 5.3.1.6 IMPROVED FEEDING OF RUMINANT LIVESTOCK

Through improving quality of feed for stall-fed or pastoral livestock, huge gains can be made in offtake weight and fertility (many pregnancies do not reach full term because of nutritional deficiencies in the mother). Feed quality from rough graze and browse, especially in drylands, is not usually of high nutritive value, nor are cut-and-carry cereal haulms following harvest. Nitrogen is invariably in short supply, which acts as a constraint on the protein synthesis needed for growth, development and fertility, and strength in a draft animal.

Palatable legume haulms, such as groundnut, in the feed mix can help correct the nitrogen deficiency, as also floor sweepings from SME flour mills grinding soybean, for example. The cake remaining from crushing sesame seed for its oil is also a valuable proteinaceous and calorific supplement, as used in Yemen and Sudan, for instance, and low-grade dates, as used in Wadi Hadramaut, Yemen.

Certain minerals also often comprise a limiting factor to growth and well-being. Mineral-lick blocks can hugely augment the milk yield of cattle, and invariably more than cover the cost of purchase. Such blocks are often present in small agro-input retailers in rural high streets in developing countries, yet because smallholders do not know of their potential value do not purchase them. Extension efforts through demonstration on-farm can change perceptions quickly in a village, once milk yields are seen to increase substantially, creating a demand thereby. There could hardly be a cheaper extension effort—the cost of a few blocks. Uptake of the "technology" by the community is immediate, by referral among relatives and neighbors. The village of Nyabushabi in south-western Uganda is a case in point, following an outreach activity from nearby Kabale University.

Some indigenous wild plants can serve as mineral-rich fodder, *Atriplex* an example of palatable feed for camels in semi-arid areas. In the Palestinian Jordan valley, one of the outcomes of movement restrictions placed on Arabs in "Area C" by Israeli authorities is that sheep and goats have become too numerous relative to the restricted grazing, which has led to massive overgrazing, reducing the sward's ability to grow back when the rains come. Owing to the high price of fodder and concentrate feed imported from Israel, and fodder grown on available arable land in Area C, flock size has had to be reduced, both of settled communities and nomadic pastoralists.

While working there during 2014, the current author became aware of a wild plant (probably family *Chenopodiaceae*), called in Arabic "ملح malh" (salt), which first spread to the Jordan Valley from Sinai in about 2009, according to local farmers. It grows on hard surfaces where other generally more competitive wild species cannot establish, yet grows into a substantial annual bush and is highly palatable for smallstock, for browse and cut-and-carry stall-feeding. The plant is highly drought-tolerant, remains green well into the dry season and seeds prolifically. These seeds could be collected once the wild plant matures and sown on rocky steep slopes where other wild plants cannot establish. As the plant is so calorific, it burns fiercely and should not be planted next to homes where accidental incineration could occur, with collateral damage. Such an initiative would reduce the amount of expensive bought-in fodder needed.

In 2011 the current author worked in greater Somalia, and in Box 5.1 offers some potential improvements which could be made to livestock feeding there.

## Box 5.1 Improved Feed for Somali Livestock

Livestock is the most important productive and export sector of Somalia, and pastoralism the main form of livelihood. It is estimated that Somalia possesses about 3.3% of the African continent's livestock, including 50% of the continent's camel population and 10% of its sheep and goats. The Somali livestock industry is mainly based on a nomadic pastoral production system, closely linked to commercial- and export-oriented marketing structures. The livestock sector creates about 65% of Somalia's job opportunities, generates about 40% of its GDP and 80% of its foreign currency earnings.

Greater Somalia's livestock sector is faced by several challenges. In addition to inability to control and eradicate trans-boundary animal diseases which lead to frequent trade bans, the Somalia livestock sector has been seriously threatened by repetitive droughts, degradation of the environment, movement restrictions, absence of sector policies, weak specialized public and private services and related institutions, dearth of specialized human resources, and the absence of processing capacity to transform and

(Continued)

**Box 5.1 (Continued)**

add value to products of animal origin, which would be an important source of employment creation. In Somaliland and Puntland, public sector institutions remain weak due to shortage of adequately trained staff and poor budgetary support to implement activities, while the Federal Government remains ineffective in many parts of South and Central Somalia due to persistent insecurity.

Livestock production and marketing from Greater Somalia are geared more toward numbers than health. Enormous value could be added to the trade through greater attention being paid to individual animal quality and offtake weight, based on the provision of improved feed. Nitrogen-rich supplementary feed could be grown under protected conditions, such as a thorn *kraal*, using irrigation sourced in *hafirs* or other water-harvesting structures. This can be provided to the stock at night when they return to the homestead. Seed of suitable crops/varieties can be availed from CGIAR stations, through the Ministries of Livestock and multiplied in-country. Additionally (following adequate applied research), there could be seeding of "medics"[25] or similar drought-tolerant prostrate leguminous genera to improve rough pasture, and as fodder supplements along seasonal pastoral migration routes or marketing corridors (eg, the Berbera Corridor). Bulked seed of these medics can be availed through countries where they are indigenous, for example Libya, or from gene banks such as SARDI or INRA[26]. Seed can be spread by airdrop, where security allows.

Use of "mineral-lick blocks" to supplement mineral deficiencies can be employed. Though these can be imported, they can also be made locally[27]. Manufacturing and distribution of these can provide employment for the private sector. Finally, better feeding is needed in port holding areas (together with proper watering and shelter to reduce stress). Just as for use of mineral blocks, such "fattening" practices have proven cost effective in other countries. The private sector is already involved in conveying fodder to such holding grounds, yet there is scope for improving the quality and quantity of this feed, not merely as subsistence rations but to enable stock to put on weight and improve their general condition just prior to export, to attain better prices.

#### 5.3.1.7 SUPPORT TO ARTISANAL FISHERIES

Fisheries and aquaculture offer ample opportunities to reduce hunger, improve nutrition, alleviate poverty and unemployment, generate economic growth and ensure better use of natural resources. However, if the current trend of unsustainable exploitation of marine resources is not reversed, the ability of the oceans to deliver food for future generations will be severely compromised.[28] Aquaculture is the world's fastest growing food production sector. Total fish supply is projected to increase to 186 million tonnes in 2030 with fisheries and aquaculture contributing equal amounts. Fisheries and aquaculture are a vital source of nutritious food and protein for billions—worldwide nearly 3 billion people receive 20% of their daily animal protein intake from fish. Employment in fisheries and aquaculture has continued to

[25] Introducing the self-seeding prostrate growth legume *Medicago sp*, to increase sward productivity and ground cover against erosion.
[26] SARDI—South Australian Research and Development Institute; INRA—*Institut National de la Recherche Agronomique*, Montpellier, France.
[27] Though such blocks may not be available in Somalia, the way to make them is described on the Kenya Agricultural Research Institute (KARI) website.
[28] Retrieved from http://www.fao.org/post-2015-mdg/14-themes/fisheries-aquaculture-oceans-seas/en/ (accessed June 11, 2015).

**FIGURE 5.2** Activity at Gyeik Taw fishing village, Andaman sea coast, Myanmar. Photo: Ulricht Schmidt.

grow faster than in agriculture—providing about 55 million jobs worldwide. Including ancillary activities (eg, processing and packaging) these sectors support the livelihoods of 10–12% of the world's population.

A relatively little-known resource on sustainable fisheries was produced by the Royal Swedish Academy of Agriculture and Forestry (KSLA) in 2009, with contributions from 52 authors. It includes a chapter stressing the importance of the artisanal subsector for employment, especially of women, in processing, marketing and other services (Finegold, 2009). While rarely involved in commercial offshore and deep-water fishing, millions of women and girls, especially in Asia (see Figs. 5.2 and 5.3), are involved in capture fisheries and aquaculture ventures. Tasks include making and repairing nets, baskets and pots, baiting hooks, setting traps and nets, fishing from small boats and canoes, and collecting seaweed, molluscs and pearls. In aquaculture, women attend to fish ponds, collect fingerlings and prawn larvae, and harvest the fish. Women play a major role in fish processing in many parts of the world, both using traditional preservation techniques and working in commercial processing plants.

The Rio+20 outcome document "The Future We Want" of June 2012 (paras 158–177) stresses the need to utilize the oceans' vast potential wealth wisely. Over recent decades, experiences and lessons learned worldwide have underlined the crucial role which artisanal and small-scale fisheries (including artisanal processing and marketing) play in general, and in developing countries in particular. By comparison with industrial and semi-industrial fisheries, they provide for sustainable livelihoods and more income and employment created per unit of investment/energy used and unit of living aquatic resource extracted, and produce less waste and discards.

**FIGURE 5.3** Fishing boats waiting to set off, Bay of Bengal coast, Thandwe district, Rakhine State, Myanmar. Photo: Ulricht Schmidt.

Artisanal and small-scale fisheries are usually owner-operated, exploit inshore waters, and use passive fishing gear such as gill nets, hooks and lines, and pots and traps which, as compared with active gear such as bottom trawls used by industrial vessels, have little if any negative impacts on the marine environment. Revenues are mostly recycled into local economies and have in many countries served as an engine of economic development.

Shallow-draft small-scale fishing boats often operate from small ports and landing sites relatively close to the fished resource. In addition to the more open areas along the coast, small-scale vessels can also exploit more restricted waters that would be difficult, even dangerous, for larger vessels. Artisanal fisheries require relatively low investment in technology and equipment and are consequentially more competitive in most developing regions where labor is cheaper than equipment. In such cases, resources within the technical reach of the small-scale sector are usually most profitably harvested, with superior returns on the capital invested as compared with industrial fishing.[29]

Commonly, artisanal fisherfolk do not engage in their craft full-time, only in the season when fish are particularly abundant. For the rest of the year, they may engage in land-based crop production or ply another artisanal trade. Being labor-intensive, artisanal fisheries are naturally suited to rural areas with high demographic growth, providing employment in catching, as well as processing and trading of fish and fishery products (see Fig. 5.4). With the constant pressure of continued population growth, migration from the hinterland and development of industrial fishing, the survival of small-scale/

[29] Retrieved from www.fao.org/fishery/topic/14753/en (accessed June 11, 2015).

**FIGURE 5.4** At the edge of Tonle Sap Great Lake, Cambodia. Continuing the value chain: artisanal processing of fish to fishpaste. Photo: Peter Degen.

artisanal fisheries depends to a large extent on the recognition and protection of traditional or acquired fishing rights.

To develop and maintain the advantages of small-scale/artisanal fisheries, the system of diversified family livelihoods characteristic of the sector should be protected and strengthened, through rights of access of the communities to a sustainable matrix of productive activities. The FAO Code of Conduct for Responsible Fisheries of 1995 recognizes their particular role in Paragraph 6.18, applauding their important contributions to employment, income and food security. States are petitioned to "appropriately protect the rights of fishers and fish workers, particularly those engaged in subsistence, small-scale and artisanal fisheries, to a secure and just livelihood, as well as preferential access, where appropriate, to traditional fishing grounds and resources in the waters under their national jurisdiction". In other paragraphs also, the Code asserts the obligations of State authorities to protect the rights of the artisanal subsector.

As well as fisheries, there are many other marine enterprises that can be run on an artisanal basis, ranging from sea cucumbers in the western Pacific and lobster capture in Vietnam, to table salad seaweeds grown on old car tyres in the sea around the Philippines. Furthermore, inland fish ponds are important in many countries to provide food and income, from mud crab culture in Indonesia and Vietnam, to the well-honed sustainable fishpond system so common in China involving chicken cages over ponds with their droppings fertilizing the water to encourage algal growth to feed the fish and ducks. Similarly, all such enterprises sustainably exploiting water bodies need protection and encouragement from State authorities, and an enabling environment for conduct of their businesses (see Section 6.6).

## 5.3.2 Food Loss and Wastage

One-third of global food production does not nourish anyone. This amount has been estimated by FAO at 1.3 billion tonnes a year (fresh weight), and is lost or wasted[30] in food *production* (lost mainly at harvesting, processing and distribution) and *consumption* systems (wastage largely with the retailer and end consumer). Most current forecasts of food availability and need ignore the possibility that measures could be taken to address wastage, assuming it will continue at the current rate (ActionAid USA, 2013) *op. cit.*

On January 22, 2013, the United Nations Environment Program (UNEP), FAO and other partner organizations such as WRAP (Waste Resources and Action Program) and governments, launched a global campaign to change the culture of waste, called "Think, Eat, Save. Reduce your Footprint".[31] In part, this campaign was stimulated by a publication of FAO (2011), and the outcome of the Rio+20 Summit in June 2012 which, in the interests of a more resource-efficient and sustainable world, promoted a 10-year framework of programs for sustainable consumption and production, part of which framework addresses the food sector.

Simple behavioral change actions along the entire food chain of production to consumption, in both developing and developed countries, by end consumers, wholesalers, retailers and the hospitality industry can significantly reduce this combined loss, in the interest of a better-fed humanity, now and in the future. In a world of over 7 billion people, increasing to more than 9 billion by 2050, wasting food cannot be defended, morally, economically or environmentally.

At the launch of the aforesaid campaign in January 2013, the UN Under-Secretary-General and UNEP Executive Director jointly lamented not only the associated waste of money involved for agricultural labor, fertilizers and opportunity cost of land and water, but also the associated greenhouse gases produced. These emanate from vehicles transporting waste food to landfills which then decomposes there.

At the same launch, the FAO Director-General said that in the developing world, around 95% of food loss and waste are unintentional losses at early stages of the food chain due to financial, managerial and technical limitations related to harvesting, storage and cooling facilities, infrastructure, packaging and marketing systems (www.fao.org/save-food/en). By contrast, in industrialized regions, almost half of the food squandered, among 300 million tonnes annually, occurs because producers, retailers and consumers discard food that is still fit for consumption (often due to quality standards which over-emphasize appearance). This itself is more than the total net production of sub-Saharan Africa, and would be sufficient to feed the more than 800 million food-insecure people then in the world.

---

[30] *Food loss* refers to food that gets spilled, spoilt or otherwise lost, or incurs reduction of quality and value, before it reaches its final product stage – at production, post-harvest, processing and distribution stages in the food supply chain. By contrast, *food waste* refers to food that successfully traverses the food supply chain to the final product stage, yet is not consumed because it is discarded, regardless of whether it has spoiled. Food waste typically occurs at retail and consumption stages in the food supply chain (www.wrap.org.uk).

[31] http://www.thinkeatsave.org.

Food wastage and loss of course have a feedback effect on the amount of food available, and hence its price and affordability, and thereby economic access to food for the world's poor. Action taken, or not taken, to bin a slightly blemished fruit or vegetable in California for instance, can indirectly influence whether someone in a developing country is hungry or not. A few examples of wastage in the author's experience will suffice here:

- The hospitality industry, no more so than in Mediterranean countries, delights in providing more food to customers than can be eaten, and the leftovers binned. The concept of bringing food as it is required rather than to swamp the table up-front is not there in many Palestinian restaurants for sure. And whilst living in Monrovia, Liberia, the author often saw small boys with bags full of collected leftovers from restaurants, heading home to the slums with their family's evening meal.
- Basic food commodities for nationals and many of the expatriate community in Kuwait are subsidized or free, so more food than is needed is collected from shops, such as pita bread, and the balance thrown away when it is a day or two old (imported wheat going to waste, creating an upward pressure on international grain prices). Currently, 1.8 million people in Kuwait (Kuwaitis and their house staff) spend on average just 5–6 KD per person per month on food, so there is little incentive to buy sparingly or to reduce waste. This costs the country a lot in "wasted" subsidies and the cost of importing the food, and certainly in the case of the sugar subsidy is partly responsible for 30% of Kuwaitis being diabetic. Previous initiatives and government statements on food consumption subsidies have not been implemented, such that the price of essential foodstuffs in Kuwait has not risen since 1965.
- The Baganda ethnicity in Uganda will never finish all the staple plantain banana on the plate, as a courtesy to show the mother or hostess that she has provided more than is required. Were the plate to be emptied, the implication is that the recipient would have eaten more had it only been provided, showing she had not done what was required of her. This "waste" will then be binned, or fed to pigs or chickens in a rural compound.

More efficient harvesting, processing, storage, transport and marketing, combined with changes in consumption behavior or government policy can result in a healthier and hunger-free world. For the campaign to achieve its potential, every player in the chain needs to be involved, from individual families and supermarkets to world leaders.[32]

## 5.4 ADDRESSING BETTER ACCESS TO, AND UTILIZATION OF, NUTRITIOUS FOOD

### 5.4.1 Employment and Job Creation Programs

A sustainable route to overcoming poverty is through creating incremental sustainable productive employment. Mainstreaming this is the gateway to assured

---

[32] Pointers toward this goal can be found at www.thinkeatsave.org and www.gov.uk/government/publications/future-of-food-and-farming.

long-term economic access to food. One way that this can come about is through fostering greater agricultural productivity and production, thereby creating more jobs on-farm, or off-farm in allied tasks, through SME development in the areas of food processing, transport of the incremental produce and selling it, agricultural mechanization/credit/extension services, construction of warehouses or road widening, and so on. In sub-Saharan Africa, for example, the agricultural sector employs 60% of the adult population, and on this score alone is a most appropriate sector to target for creating incremental employment.

This new wealth will be created only if agribusiness activities are demand-/market-led. Markets themselves need improving, and this can be done even at a basic level—well within living memory in the lower Himals of eastern Nepal is the initiative of one person who started the weekly "*haat bazaar*" system, such an important economic and social institution nowadays, where goods are sold or bartered. In Section 4.5.3 another example is given from 2012 of the market in Ganta township of Nimba country, Liberia, being upgraded on the initiative of government following a USAID-funded project identification exercise. Ganta market serves as a large commercial hub not just for northern Liberians, but for people across the border in Guinea and Ivory Coast.

New wealth may also be created in related SMEs which can benefit from the new agricultural wealth—SMEs that make shoes or bicycles, which the newly "rich" agricultural community can buy.

A way to bump-start an invigorated economy is through "job creation schemes", meaning the provision of new opportunities for paid employment, especially for those who are unemployed (see Section 4.3). A public works scheme, suggested by economist Maynard Keynes and espoused and funded by government, played a major role in getting the United States out of its depression following the 1929 stock market crash. International institutions and donors can undertake such schemes too, the PSNP in Ethiopia having been mentioned elsewhere (see Case Study 1 on the book's companion website). Such schemes are suitable for those individuals or families who have able-bodied individuals who can work, but not for those who are unable, or no longer able—the elderly and infirm, severely disabled and infants. These need a system which provides humanitarian assistance, if not provided for by the earnings of those in paid employment—there are many in work who are not paid for their labor, family women in developing countries being prime examples (see Section 6.2).

In addition to PSNP, another example is the Job Creation Project in the Gaza Strip of Palestine for instance, implemented by the Office of the President and funded by the Spanish Federal and Regional governments. This provides short-term Cash for Work tasks in the field of agriculture, whereby Gazans can earn a little to support themselves and their families instead of needing to rely on humanitarian aid through WFP, UNRWA or charities. Another job creation initiative in Gaza, the Economic Recovery Program, run by Oxfam and funded by Danida, supports key economic growth sectors (agriculture, dairy, fisheries Information and Communications Technology (ICT)), "creating spaces" for promising small-scale enterprises, many run by women, to add value.

### 5.4.2 Public Policies for Social Protection

Social protection starts at the level of families where a member who is disadvantaged through young or old age and/or infirmity and unable to care for him/herself is looked after by another member(s) of that immediate or extended family who is/are somewhat better-endowed. There have always been individuals who have fallen through this kinship safety net though, giving rise to loss of all forms of security and dignity, perhaps adequate shelter and companionship, and vulnerable to hunger, undernutrition, and poor health even unto death. Protection of these individuals could enable them to stay above the threshold of mining their assets beyond that critical level in order to survive that it becomes impossible to recover—like selling the thatch on the family house to buy food (see Section 4.1).

In enlightened nations, governments assume responsibility for providing a degree of "social protection" against poverty, and recurrent vulnerability to, and onset of, food and nutrition insecurity, as a discretionary duty of care. More recently, this has been upgraded to a "rights issue" obligation, the human right to nutritious food on a daily basis and ability to lead a productive, healthy and dignified life, this right enshrined in a legally empowered social protection system. A rights-based approach is both morally and legally judicious, and likely to result in improved food security outcomes. Access to social protection and food security are universal human rights, recognized by most States, and need to be written into national laws if not already there (as in the amended Basic Law of Palestine of 2003, and its National Food Security Strategy of 2005).

Social protection measures currently available exhibit a range of forms and combinations, through the formulation of policies and their implementation, involving various stakeholder groups. The degree of success and equitable coverage of implementation vary from country to country and within-country, dependent on the amount of largesse available for distribution, and degree to which this is (unfortunately) influenced by ethnicity and politics.

In the developing world, the concept of social protection has received increased attention since the start of the new "Millennium". By then no clear consensus had emerged concerning design and implementation modalities of the three instruments of protection policy—*social assistance* (payments and/or in-kind transfers to support and enable the poor), *social insurance* (permanent protection against risk and adversity), and *social inclusion* (enhancing the capability of the marginalized to participate fully in economic and social life, and to access social services).

In 2010, the CFS, complying with the mandate accorded it by the UN, requested a High Level Panel of Experts (HLPE) to work on Social Protection, and more specifically on ways to lessen vulnerability through social and productive safety net programs and policies with respect to food and nutritional security, taking into consideration differing conditions across countries and regions. CFS further required that the study should include a review of the impact of existing policies on the improvement of livelihoods and local production, living conditions, nutrition and resilience of vulnerable populations, especially small-scale rural producers, urban and rural poor, women and children.

The resulting 100 page Report of the HLPE, containing its analysis, conclusions and recommendations was availed to the CFS in 2012 (HLPE, 2012). A summary of the Report's observations, and recommendations for using instruments of social protection more effectively is presented below, and the reader is encouraged to consult that Report for more detail (together with Case Study 1 on the book's companion website), about the "twin-track" approach of the PSNP in Ethiopia, which combines short-term assistance with long-term investment in livelihoods.

#### 5.4.2.1 SOME KEY POINTS OF THE HLPE REPORT

1. The appropriate social protection response to chronic poverty-related food insecurity is *social assistance* linked to livelihood promotion measures that enhance incomes, and therefore access to food. If well-designed, social protection systems are not "deadweight" burdens on fiscal systems. By preventing the depletion of assets and reducing the personal risk to the poor of investing in their livelihoods, social protection can be a "win-win" strategy—pro-poor and pro-growth.
2. The various causes of food insecurity can be addressed through an array of social protection policy responses. For instance, food production by smallholders may be boosted with *input subsidies*, while harvest failures or livestock losses can be compensated with *agricultural insurance*. Unemployment or under-employment can be addressed through *public works programs*. Restricted market access to food can be addressed on the demand side (*food price stabilization and price subsidies*) or on the supply side (*grain reserves*). Inadequate access to food can also be addressed directly, through transfers of food (*food supplements*, *complementary food demonstrations and school feeding*) or *cash transfers* (conditional or unconditional). Complementing such interventions with job creation and on-farm research-extension services, for example through improving both access and availability aspects of food security, should offer more permanence and sustainability to the earlier-mentioned mechanisms. Several of the instruments cited earlier have shown promise when combined, examples cited being "Challenging the Frontiers of Poverty Reduction Program" implemented by the NGO Brac in Bangladesh, the flagship social protection "Vision 2020 Umurenge Program" in Rwanda, and the "Zero Hunger" program in Brazil, launched in 2003 as part of the National System and Policy for Food and Nutrition Security.
3. While nationally owned social protection programs for food security is the ideal, many governments will require technical and financial support

from UN organizations and other development partners, at least for an interim period.

4. Several challenges arise in designing and implementing social protection programs: for instance, how best to target deserving beneficiaries, avoid "dependency", ensure accountability for funds and equitable disbursement, and scale-up proven interventions to other areas of the country or other countries.
5. Social protection initiatives should be gender-sensitive, one reason being that it is the womenfolk who both provide care for children in the home *and* grow the food crops (and they will likely make better decisions than men on how to dispose of payments in cash or vouchers in support of food and nutrition security—see Section 6.2). Program design should consider the tension between a woman's role in care-giving and her role as income earner in public works programs.
6. The HLPE Report recommends that the CFS could play a role in ensuring that rigorous and credible *evaluations* of social protection initiatives are conducted. Collation of these should provide a global lesson-learning opportunity, to further the observations in the HLPE Report itself. The CFS should also ensure that the recommendations in the Report are incorporated into the final version of the *Global Strategic Framework for Food Security and Nutrition* (GSF), which CFS is developing.[33]
7. Some families for social protection are able to "graduate" out of a scheme, when their condition improves (because of a good rainy season, say); these would likely have the capacity to provide "labor". Other families or individuals have limited or no capacity to graduate from a scheme (see Case Study 1 on the book's companion website).

In addition to what has been presented earlier from the HLPE Report of 2012, other types of economic and social policies and strategies to improve food access (availability, consumption and stability) could also be considered, such as those relating to trade and tax (domestic, regional and international), which can promote or hinder economic development/food security. These would involve global policies, for example, those espoused by the World Trade Organization (WTO) and the EU, and cross-border and regional cooperation agreements. An example of the latter is those on nutrition, livestock disease surveillance and climate change between Israel and Palestine, which are cordial and technical, free of political spin. Space does not allow an examination of these policies and agreements here.

---

[33] The purpose of this Framework, the second version of which was ready in October 2013, is to improve coordination and guide synchronized action by a wide range of stakeholders in support of global, regional and country-led actions to prevent future food crises, eliminate hunger and ensure food and nutrition security for us all. The GSF offers guidelines and recommendations for coherent action at the global, regional, and country levels by the full range of stakeholders, while emphasizing the central role of national ownership of programs.

## 5.5 ADDRESSING BOTH AVAILABILITY AND ACCESS TOGETHER

### 5.5.1 Food Reserves

#### 5.5.1.1 INTRODUCTION

Harvests are unpredictable, and food consumption is neither elastic nor optional. Stockpiling food in times of plenty to guard against famine if the harvest fails has been a self-evident strategy for humankind, at homestead and local community level. The commodities that need storing reflect food consumption norms and preferences.

With regard to large-scale storage, not all commodities are stored in bulk for a long time, such as fresh milk; food grains require silos or warehouses in which to store sacks, while meat, fruit and vegetables require refrigerated facilities. All require power sources for microclimate control, with grid electricity or generator fuel supplies sometimes threatened by shortfalls. Box 5.2 indicates an overview of the perilous dynamics in global grain stock inventories.

#### 5.5.1.2 STRATEGIC FOOD RESERVES

Strategic food reserves under the control of rulers of the time were kept in ancient Egypt (since 1750 BC), China (since AD 498) and during the Roman Empire. Nowadays, governments in countries such as Brazil, India, China,

**Box 5.2 Global Grain Stocks**

Failing grain harvests in the United States, Ukraine and other countries in 2012, eroded global food reserves to their lowest level since 1974. The United States, which in 2012 experienced record heat waves and droughts, at the end of that year held in reserve an historically low 6.5% of the corn (maize) that the country was expected to consume in 2013. With global food consumption exceeding the amount grown for 6 of the 11 years prior to 2012, countries ran down grain reserves from an average of 107 days of consumption in 2002 to under 74 days in 2012.

Prices of main food crops such as wheat and corn have risen close to those that sparked riots in 25 countries in 2008. In 2014 FAO estimated that at least 805 million people were chronically undernourished, and the food crisis is growing in the "Middle East" and Africa. Wheat production in 2012 was expected to be 5.2% below 2011, with yields of most other crops, except rice, also falling. The area of wheat and barley being grown is not increasing in response to this shortage. In October 2012, Oxfam predicted that the price of key staples, including wheat and rice, may double over the forthcoming 20 years as a result of shortfalls of grain, threatening disastrous consequences for poor people who spend a large proportion of their income on food. The United Nations has warned that this could precipitate a major global hunger crisis imminently, with low-income developing countries least able to ride such a crisis.

"Middle East" countries are going to be huge importers of food no matter what, many but not all being able to afford the escalating world price of grain. Arab countries together are the largest net importers of cereal calories in the world, importing roughly 56% of the cereal calories they consume. Given the limited resources of fresh water and arable land in Arab countries, wheat imports may rise by almost 75% over the next 30 years (World Bank, 2012).

Indonesia, Mali, Senegal, Zambia and Malawi all maintain varying types of national food reserves (ActionAid, 2011).

There are several potential purposes of strategic food reserves nowadays (Murphy, 2009):

1. To prepare for food emergencies.
2. To smooth out volatile prices.
3. To correct any failure of commercial food markets.
4. To complement or replace the private sector.

The four purposes are not mutually exclusive, especially the first two which will often support each other. Strategic food reserves under public governance may be at sub-national, national, regional and/or global level. Box 5.3 provides examples of regional grain reserves.

Humanitarian assistance food stocks may conceptually be distinguished from other public food security stocks on the basis that the former specifically target vulnerable groups, whereas the latter are directed toward overall availability and prices in local markets (in practice, the two may be integrated).

Even countries which are not categorized as "developing", yet are nevertheless exposed (from regional conflict for instance) to high risk to their trade and food security on which it is based, can benefit from strategic reserves, such as Persian Gulf States. Kuwait, high in the 2014 UNDP Human Development Index at number 46 is an example (see Preface). Such countries may be rich with petrodollars but due to negligible rainfall and dwindling aquifer levels have insufficient water to produce their own food crops, apart from high-value vegetables and fruits, and a basic livestock industry using imported feed.

Commercial entities which need to run at a profit cannot be expected to provide long-term *strategic* reserves for the public good or be mindful, as governments should be, of using those reserves as a tool to prevent speculative hoarding and dampen commodity price rises.

**Box 5.3 Regional Food Reserves**

The Association of South-East Asian Nations (ASEAN) has had a regional food reserve for more than 35 years and recently reinforced its policy for it. It originally launched the ASEAN Food Security Reserve (AFSR) in 1979 and is now replacing it with an ASEAN Plus Three Emergency Rice Reserve (APTERR), which includes China, Japan and South Korea as members in addition to the ten members of ASEAN (UNCTAD, 2011).

There are also several ongoing efforts to establish regional food reserves in Asia, Africa and Latin America. For instance, the Latin American and Caribbean Emergency Preparedness and Response Network (LACERN) has partnered with WFP to set up an effective regional emergency food reserve to respond to natural disasters such as droughts, floods, hurricanes and earthquakes. The Network has a main hub in Panama City, plus three subregional hubs. A regional food reserve has also been created in West Africa, and plans mooted in 2015 of establishing a similar one in southern Africa.

##### 5.5.1.2.1 Summary of Potential Advantages of Strategic Food Reserves

1. *Enhances food availability:* Large-scale storage of essential food commodities in a country or region can serve as a buffer facility, in the event of an emergency which threatens widespread food insecurity, rather than relying on markets which can fail as they did in 2007–08. Sometimes a cash reserve may be tied to a physical stock facility to allow emergency purchase of stock from outside of the crisis area concerned. These *strategic* reserves fulfill a risk-mitigating public service, a short-term management tool, with central or local government either controlling them or contracting out that function to a third party. Such reserves are separate from any food commodity storage which is part of the commercial private sector.

   The need and potential benefit for such strategic reserves is proportional to the likelihood of emergencies arising which would affect local food production and availability, from conflict, flood, drought, earthquake, crop pest and disease outbreaks, for instance; proportional also to the country's ability to access, pay for and efficiently distribute essential food commodities sourced outside the nation's borders. This latter composite necessity is particularly challenging for landlocked countries, dependent on the goodwill of their neighbors and efficient management of their seaport facilities; also on the ability of road or railway infrastructure of the crisis-affected country and that linking it to the seaport to withstand the extra demand and use, and the availability of reliable rolling stock.

   For example, Rwanda in east-central Africa was faced with conflict-related food security challenges during the 1990s requiring massive road shipment of grain across intervening countries from Indian Ocean ports (which damaged those countries' roads). Since then the Rwandan government has developed a Strategic Grain Reserve concept comprising storage facilities together with a cash reserve for procurement of stock, has liberalized grain markets and encouraged private sector involvement in improving value chains and postharvest investment (Government of Rwanda, 2011).

   FAO (2009a,b) noted a growing interest in grain reserves at *local* as well as national levels, citing Burkina Faso, Comoros, DR Congo, Madagascar, Malawi, Zambia, Nicaragua and Pakistan as countries with proposals to strengthen existing grain reserves or to introduce them. Some African countries, including Burkina Faso, Burundi and The Gambia, have focused on building village-level grain reserves to ensure food security at community level. The Asian Partnership for the Development of Human Resources in Rural Asia (AsiaDHRRA) promotes community reserves as an important element of its efforts to strengthen local food systems. Community reserves, under local control, are immediately accessible to the local population, comprising local products so that dietary habits are preserved, and dependency on staples from outside the community is reduced.

2. *Improved economic access to food:* Through smoothing price volatility and reducing spikes, economic access to food is improved. During the 2007–08 FPC, countries such as Bangladesh, Indonesia, China and India had sufficiently large food reserves and public distribution systems to stabilize prices in domestic markets. While some smaller countries such as Malawi had re-built their public food reserves so they could manage and release public stocks and protect themselves during that food crisis, many other low-income food-deficit countries found that the insignificant size of their reserves allowed them only to render a safety net function during the crisis, where stocks were used for distribution or subsidized sales to the vulnerable, with little impact on prices (Mousseau, 2010). *Where present, reserves should be large enough to be used for both price-control and emergency food security*. An exemplary national food reserve system is that in Brazil which uses its national food reserve system to stabilize local prices of staple crops such as corn, and to support smallholder agriculture by purchasing from it.
3. *Support to producers:* To build up stocks, governments or their agents can purchase grain or other food commodities at harvest when prices are relatively low (thereby helping to raise prices to support local producers). These stocks can be released when prices are higher, in the traditional "hungry period" after planting but before harvest, this helping to relieve the open market shortage and upward pressure on prices. Many strategic reserves use adjustable "price-band" mechanisms which trigger action at minimum and maximum target price levels. In this way, reserves can help protect farmers' incomes and create more certainty for making crop planting decisions, as well as mitigating the impact of price rises on poor consumers.
4. *Underpin food entitlement schemes:* Food reserves enable public distribution schemes or Food-for-Work schemes, through providing subsidized staple foods for the rural and urban poor. A model of propriety is the Ethiopian Food Security Reserve, managed by an autonomous administration which has proven its effectiveness on several occasions since the 1990s (IFPRI, 2011). The maximum stock level is maintained at 407,000 tonnes, and stocks are released to distribution agents in a national donor-funded safety net program, with the Ethiopian government paying the running costs.

#### 5.5.1.2.2 Potential Disadvantages of Strategic Food Reserves

1. *Cost:* Public sector grain reserves and associated distribution can be expensive to set up and run.
2. *Management:* Excellent management is needed, to procure and maintain stock in good condition, to rotate and distribute it. There are instances in which poor management has wasted money and the reserve been a hindrance in times of crisis, such as that in Malawi in 2002. By the 1980s, food reserves were being heavily criticized, the producer benefits of food

reserve management often captured by better-off farmers at the expense of the net food-consuming rural poor. Food reserve agencies were criticized for acting as monopoly buyers and sellers of grain, and manipulating food prices by intervening in the market, which was considered inefficient and expensive, as well as encouraging corrupt practice and preventing the emergence of a competitive private grain sector. In low-income contexts with weak infrastructure, the transaction costs of trading in deep rural areas were often very high, and demand flagged by poor consumers was too weak to attract traders. Consequently, when governments finally withdrew, a vacuum was left which traders did not fill, and the rural poor were left more vulnerable than before, at least for a transitional period.

3. *Market distortion:* Potential market distortion causes concern to the WTO. The distortion can also be seen as an advantage though, as reserves act as an intervention to compensate for what markets cannot achieve, and smooth out price volatility. This may, however, depress some private sector activity, or incentivize and lock farmers into growing certain crops, such as rice and wheat which are held in the reserve, rather than diversifying into other crops (IFPRI, 2006).

#### 5.5.1.3 THE WISDOM OF PUBLIC SECTOR FOOD RESERVES

As a result of the disparate perspectives cited earlier, what is the best way forward ? Despite their cost and management challenges, food reserves have proved an important food security policy instrument. Furthermore, innovative mechanisms for reducing costs and improving efficiency are being piloted, such as splitting the reserve between a physical stock and a financial fund, or using futures markets to source food supplies promptly through hedging arrangements.

ActionAid (2011) *op. cit.* and Murphy (2009) *op. cit.* note that concerns about high running costs, market distortions, corruption, rent seeking, political interference and inefficiency mean that a large number of national food reserves have been scaled back or dismantled since 1990. Instead, poor countries have been encouraged to rely on imports from global markets for their food requirements.

Following an earlier attempt in the World Food Conference of 1974, lent support by US President Ford, which went nowhere, the national/international public grain reserves issue is receiving support again. At the G-8 summit, held in L'Aquila, Italy in July 2009, the gathered Heads of State signed a declaration stating "The feasibility, effectiveness and administrative modalities of a system of stockholding in dealing with humanitarian food emergencies or as a means to limit price volatility need to be further explored. We call upon the relevant International Institutions to provide us with evidence allowing us to make responsible strategic choices on this specific issue".

The literature on public grain reserves ranges from trade vested interests, which have little or no place for public food reserves, to those which stress their importance—NGO's such as Acord (Acord, 2014) and ActionAid, the UN Conference on Trade and Development (UNCTAD, 2011) FAO and the former UN Special Rapporteur on the Right to Food, Olivier De Schutter (De Schutter, 2013).

On the eve of a high-level WTO summit in Bali, Indonesia in December 2013, which endeavored to reach agreement on proposals on developing countries' food stockholding for food security, as part of the Doha Round trade negotiations, legal expert De Schutter said "Food reserves are a crucial tool, not just in humanitarian crises, but in the everyday struggle to provide stable income to farmers and to ensure a steady flow of affordable foodstuffs for poor consumers, many of whom lack a basic social safety net". Calling for developing countries to be granted the freedom to use public food reserves to help secure the right to food, without the threat of sanctions under WTO rules, he said "Trade rules must be shaped around the food security policies that developing countries need, rather than policies having to tiptoe around WTO rules". The UN Special Rapporteur warned that food security is at high risk when countries become overly dependent on global markets, as shown during the global food crisis of 2007–08. As Fritz (2014) also points out "Enabling food security programs involving price support, grain reserves and marketing boards basically requires a reframing of WTO rules. The purpose of these rules must be to facilitate food security policies, not constraining them, even if this means a potential barrier to trade".

UNCTAD points out that the food price crisis since 2007–08 has drawn new attention to the problems of food supply in poor countries, as well as the hollowness of previous advice to them to rely on world markets to close gaps in supply. FAO (2009a,b) in their papers on country responses to the food security crisis states that many existing national food reserves worked well to protect millions of poor people by releasing public stocks during the 2007–08 FPC, and a variety of national food reserves are proving to be flexible, effective and progressive public policy tools. The management and release of public stocks, often coupled with subsidized sales of food, was a key response to high prices during 2007–08. Stock interventions took place in at least 35 countries during the crisis, including Burkina Faso, Cameroon, Ethiopia, Senegal, Kenya, Nigeria, Cambodia, India, China and Pakistan. FAO says that those countries with reserve stocks were "able to respond more quickly and cheaply than those with limited or no reserves".

As Gilbert (2011) argues, the food security debate is often posed as a choice between trade and food stocks, but this is misleading since the two strategies can be complementary. Countries need to achieve a balanced food security policy. Asian rice-producing and -consuming countries have largely managed to achieve a good balance between trade and stocks, through relatively light government interventions and procurements allowing an efficient private sector to prosper. By contrast, in many developing countries in Africa, the formal grains sector is dominated by governments, WFP and other agencies, as African food markets are expected to function poorly.

To address market failures and price volatility at the regional, national, and local level, ActionAid presses the G20 to learn the painful lessons of the 2007–08 food crisis, and in international fora support the crucial preventative role of strategic food reserves. Their establishment, maintenance and functionality to prevent potential death and disruption to economies and livelihoods costs less to achieve than the massive humanitarian efforts needed to address a future food shortage crisis. FAO produced a manual on strategic food reserves in 1997 (FAO, 1997). The need has never been greater, with food prices as high in 2014 as they were in the 2008 crisis, world stocks of grain lower than then and millions of people still hungry and undernourished, or vulnerable to succumbing. However, while there is a case for a strategic reserve to ensure that emergency food supplies are readily available, given their poor track record globally, national-level food reserves should be used to complement rather than substitute for other (sometimes more effective) social transfer interventions.

The European Commission, in the Terms of Reference for a global study (*EC ToRs of a global food reserve assignment*)[34] it mounted in 2015 to identify how best food reserves may complement and be integrated with social transfers, both nationally and regionally, cited some important questions to be addressed in order to maximize synergies, namely:

1. The supply of in-kind transfers out of the food reserve in response to food crises (how appropriate are food transfers, rather than cash? In what situations and circumstances? How to resolve targeting issues? Delivery mechanisms? Comparative efficiency?)
2. The use of local purchase into the food reserve to guarantee prices to (poor) local producers, thereby increasing their income and boosting local agricultural production (issues of seasonality; selection of commodities; price setting mechanisms).
3. The release of stocks out of the food reserve for price stabilization and to ensure affordability of staples by (poor) consumers (issues of transparency and coherence with private traders; cost-effectiveness compared with direct cash transfers).

#### 5.5.1.3.1 Integration of Food Reserves with Social Transfers

Food reserves can ensure that supplies of emergency food aid are readily available after a crisis. Yet food reserve management can also potentially complement other more developmental social transfer instruments that improve national and household food security. They achieve this by addressing problems of market access to food through both the demand side (supporting food price stabilization) and the supply side (ensuring availability).

The standard recommendation given to governments is to let domestic prices adjust to international prices. But international prices have become (and are likely to remain) highly volatile. And because food represents a relatively large share of developing countries' consumption basket, rapidly rising food

[34] Approval to quote was provided by the EC, on July 23, 2015.

prices cause inflationary pressure, hurt the living standards of many of the poor and near-poor (between them, often a country's majority), and may also trigger widespread social discontent. To avoid such outcomes, governments often resort to a variety of interventions designed to drive a wedge between international and domestic food prices: releasing buffer stocks or emergency food reserves is among the least distortionary of the options available.

Food reserves typically operate by defending a "price band" for staple cereals—a floor price for producers and a ceiling price for consumers—which they do by buying from farmers at a gazetted price, and managing market supplies to prevent excessive price rises. After the annual harvest, the food reserve agency either purchases surplus produce from local farmers or imports food and stores it in national grain banks. Later in the year, when on-farm granaries are depleted and market prices start rising, the agency releases some of this stock onto the market at cost price (plus storage), to stabilize supplies and dampen price rises. In drought or other bad years, the government might distribute the stored grain as food aid.

From a social protection perspective, food reserves are theoretically a "win-win" scenario—a win for poor farmers who know they can sell their output at a guaranteed price; a win for poor consumers who are no longer exposed to damaging price spikes; and a win for governments because (a) the prices set for producers and consumers are such that the reserve, unlike food subsidies, should operate at little cost to the state; and (b) if an emergency response is needed to a shock, the government has physical stocks it can distribute immediately. However, the appropriate form, level, and financing of food reserves require careful planning, and their management—procurement, storage, and release protocols—requires constant vigilance.

#### 5.5.1.4 TECHNICAL AND OPERATIONAL ASPECTS OF GRAIN STORAGE

To ensure that the cereal and pulse grains, vegetable oil, sugar, powdered milk, red and white meat, eggs and any other essential food commodities are stored safely to avoid deterioration, requires detailed planning and excellent operational and financial management. Such planning is needed to *conceptualize* the strategic food stock facility in the first place, in relation to what may already exist, in quantitative and qualitative terms, and to costs—for every dollar spent managing strategic reserves, there is one dollar less to invest in other key areas such as education and health.

The *set-up plan* must be grounded in terms of technical feasibility, a high "doability" and relatively low risk assessment (and profitability if the grain store is commercial). This includes location of the food reserve (in-country and/or in another country as a national or regional hub and spoke model under which wheat, for example, might be delivered to one country and then divided up for shipment to other countries in the region); distribution logistics; size (how many months' or years' supply should be in the reserve); role of the private sector; foresight (using "gap analysis" assessment) on how consumption demand will change over time in relation to population increase and any local production; and how resilient is it to local shocks—namely, how to be sure that the reserve food would be available and mobilized when needed because of crisis "at home", neither be subject to annulment because of regional power play, nor be hostage to political or climatic crises elsewhere in the world.

Ideally, national food reserves should be integrated into wider rural development and food security strategies (promoting local production and involving smallholder farmers and CSOs in deciding where the store should be sited), and in their governance structures, and this is increasingly happening and being deemed best practice, judged by its results. Such consultation requires understanding the local politics, so that one group is not favored over another, sowing discontent. For all this to happen, accessing local knowledge is paramount, with due attention to gender and sustainability dimensions.

To assure best practices in *recurrent operational maintenance and management*, there needs to be a clear mandate, with authority and finances to fulfill it in ways deemed best under a given set of circumstances; transparent, efficient and accountable procurement, distribution and anti-corruption systems in place; and, the whole being apolitical and having independent oversight. Grain silos facilitate first in-first out usage so that produce is subject to a turnover in supply, such that the oldest is milled or otherwise processed and/or marketed to make room for fresh supplies. The relationship between levels of stocking, use and replenishment needs to be perfect to maintain the integrity of the reserve, and keep spoilage to a minimum. Microclimate control is mandatory. Safety concerns include reducing the chance of dust explosions.[35] Each commodity has different requirements—rice needs "aging" for 6 months before being released, for instance, and grain from Canada and Australia will likely arrive at the distant storage depot having different moisture contents, and need drying to different extents. Related government policies and regulations need compliance, and so on.

### 5.5.2 Better Identification of Investment and Trade Opportunities

As observed by the Committee on World Food Security (CFS) in the First Draft Principles (FAO, 2014), there is significant evidence that investing in agriculture and food systems is one of the most effective ways of reducing hunger and poverty, through building stronger and more resilient communities. Unfortunately, the very regions where food insecurity and poverty are most widespread are the regions where agricultural investment has been stagnant or declining. Addressing the dimensions of food security and nutrition—availability, access, stability, and utilization—requires a significant increase

[35] This being cited as the risk of which any grain silo operator was most concerned to mitigate, when the current author visited the silos of the Kuwait Flour Mills and Bakeries in Kuwait city, in November 2013.

in responsible investment in agriculture and food systems. Such investments, including agricultural infrastructure, value chains (Schaffnit-Chatterjee, 2014), research, education, extension and other relevant services, can be carried out by a multitude of actors including public and private sectors (sometimes as a Public-Private Partnership—see Section 3.6.2), domestic and foreign investors and donor partners.

Larger blocks of land, which can be worked with greater effectiveness using economies of scale, can lead to enhanced productivity and profitability, if run on a commercial basis. It is well-established that small farming units is one of the factors of production that has to change to enable this, in countries of Central Asia, for instance, where following the collapse of the Soviet Union in 1991, huge State farms were broken up into small land areas for distribution to the people. Otherwise the drudgery and poverty associated with subsistence farming will unlikely be conquered. There are options available, through cooperatives for example, which would not impinge on households' independence. No one needs to force a change; it will be for the household to choose when offered an alternative to move into the semi-commercial realm of farming, as has been successfully accomplished in Jenin governorate, West Bank of Palestine for instance, through inward investment by local entrepreneurs. Building agricultural processing factories to add value to primary produce has created local markets for the latter, which are availed to local consumers or exported to Jordan and further afield. The Al-Naser factory is an example, whereas other enterprises such as Canaan Fair Trade for "organic" produce have involved institutional investment to supplement the local.

Further to the need for investment discussed in Section 1.7, an example of donor funding to bump-start that process, in Liberia, is recorded in Section 4.5.3. During the study cited, private entrepreneur funding sources were identified in plenty, foreign funds already deposited through a lawyer's office in the capital Monrovia. The study was able to match these funds with agricultural opportunities. Other funds were identified on the web, with business interests in the United States seeking opportunities to invest in Liberia. Efforts are under way to reduce the red tape within the public sector to make the conduct of business less burdensome, following which the Chamber of Commerce will become more active. Those who think western Africa equates only with the Sahara need to visit Liberia to see the abundant rain in season, sufficient to support tropical tree crops, including the largest single block rubber plantation in the world, and how easily Liberia could once again become a breadbasket of Africa as well as its resurgence as the regional trading hub it used to be.

The marketing of crude rubber in Liberia is exemplary, drawing in smallholders as outgrowers to large concessions. Buying points on rural roads are well-marked and well-known, and the day's price advertised there on large boards. Its marketing matches that of the *khat* crop in Yemen, one of the relatively few things in that country which works well, on a timely demand-led basis.

Investment in value chains and marketing, and the associated transportation networks and rolling stock, has great potential in the developing world, yet market economics require that smallholders themselves need to become well-organized, to pool their efforts—no lorry is going to venture from Monrovia port to a village in Nimba county to collect just one tonne of fermented cocoa beans. The beans must be bulked at village level from various farms and stored well in a warehouse, to make it worthwhile for the haulier to come from the south and for that haulier to pay well. Otherwise up to half of Liberian-grown cocoa beans will continue to be "smuggled" in sacks on motorbikes across the northern border into Ivory Coast or Guinea, where currently cocoa marketing is far better organized. Revenues to the Liberian government are going a-begging because of this internal marketing dysfunction. The disparity between the efficiency of the rubber and cocoa agro-industries in Liberia is startling.

International agribusiness investment networking is well advanced in some countries, and developing nations need to better exploit opportunities therein. As one example, Global AgInvesting (GAI), based in the United States, considers itself the world's leading resource for events, research, and insight into the global agricultural investment sector. Its website says that since 2009, GAI conferences continue to be the most critical and well-attended agriculture investment events in the world. GAI Research & Insight, launched April 23, 2012, is a web platform providing news and analysis, editorials, research pieces and a directory of major players. In the year from June 2015 to April 2016, GAI and its associated agribusiness consultancy partner plan to host major investment conferences in San Francisco, New York, London, Singapore and Dubai.

China comprises a fast-expanding agricultural market, due to its size, its population and strong economic performance. Since the government's outreach policy initiatives dating from 2004, China requires ready access to best international practices and standards to sustain its growth and proven ability to lift its masses out of poverty. These it can source in the mature research and development global markets on its doorstep, such as in New Zealand (dairy) and Australia (grain and beef). For instance, to reach its target of better self-sufficiency in rice, wheat and corn production, China needs superior seeds, fertilizers and agrochemicals to optimise crop productivity, production and protection. Provision of these technologies, through training and supply of quality physical inputs require investments both from within and outside China in order to enable the Green Revolution in China to continue.

Similarly, South America is a leading global agricultural market, accounting for approximately 10% of global agricultural product export. The continent has many commercial farms, for instance in Brazil, Argentina, Chile, Uruguay and Colombia. Corn and wheat are the main grains consumed in most of the continent. The main commercial cattle-raising centers comprise the fertile lowland (*Pampa*) of Brazil, Argentina and Uruguay and the tropical grassland plains (*Los Llanos*) of Colombia and Venezuela. Around 20% of the world's beef comes from South America.

Since the 1990s, and particularly so most recently, South America has experienced considerable growth in agricultural production, through both technical innovation allowing increased productivity and extended land use, and increased market demand—especially in the high end market and for meat products. The continent's agricultural renaissance may be ascribed to massive investment by the local and international private sector, development banks and governments, in research and application, infrastructure and transport. This is particularly so in Brazil, where agriculture accounts for around 8% of GDP and employs a quarter of the labor force, with the country being the world's third largest producer of corn. Between 1960 and 2011, Brazil increased its grain production by 774%, and livestock production by 251%. Livestock in Brazil covers 170 million hectares of production, with exports to 180 countries.[36] Clearly Brazil and other countries on the South American continent are major players in assuring food availability and food security for millions of people beyond their borders.

The role of governments is crucial as an enabler to this investment, encouraging market-led enterprise, with the necessary regulations to prevent cartels from manipulating food markets (as apparently happened for rice in Liberia, during the 2008 FPC), yet not so much regulation that investment and trade are inhibited.

The funding of trade-related projects is becoming increasingly common by some donors, such as USAID, the UK government's Department for International Development (DfID) and the EU, promoting value chain analysis, local food processing and profitability. Investment is also being made in programs such as the Productive Safety Net Program (PSNP) in Ethiopia (see Case Study 1 on the book's companion website), led by the World Bank, in which public works resculpt the environment to make the land more productive while providing Cash for Work for thousands of citizens, and programs to improve resilience against food and nutrition insecurity (see Sections 5.1.3 and 6.5).

#### 5.5.2.1 INTER-AGENCY WORKING GROUP (IAWG) ON PRIVATE INVESTMENT AND JOB CREATION

In November 2010, G-20 leaders adopted the Seoul Summit Declaration, recognizing "*the critical role of the private sector to create jobs and wealth, and the need for a policy environment that supports sustainable private sector-led investment and growth*" as one of the six core principles of the Seoul Consensus and the Multi-Year Action Plan on Development (MYAP). G-20 leaders committed to "*identify, enhance and promote responsible private investment in value chains and develop key indicators for measuring and maximizing the economic and employment impact of private sector investment*". An Inter-Agency Working Group (IAWG) on the Private Investment and Job Creation Pillar of the G20 Multi-Year Action Plan on Development, led by UNCTAD, was formed to support the G20 to achieve this objective.

In collaboration with local stakeholders, for the G20 Summit in Cannes in 2011, IAWG submitted an interim report which introduced the "indicator

[36] Retrieved from www.agriacorp.com/about.asp??id = 32 (accessed May 28, 2015).

framework" as a policy tool to measure, and thereby maximize, economic value-added and job creation arising from private sector investment in value chains. The indicators framework methodology can serve as a critical first step for governments to identify potential actions on sector-specific investment policy making, although further analysis would be required for drawing more concrete policy guidelines. At the G20 Summit in Los Cabos in 2012, IAWG presented the "Final Report on Private Investment and Job Creation" which made policy recommendations for improving the business climate and regulatory framework for foreign and domestic investment, and assisting developing countries to attract the most value-adding investment to their economies.

Subsequent to the presentation and approval of the "indicator framework", UNCTAD tested the proposed methodology in six low-income countries (Bangladesh, Cambodia, Dominican Republic, Lao PDR, Mongolia and Mozambique) across five sectors (agribusiness being one), with the objective of developing specific policy recommendations for those countries, and also to use the experience to refine the indicator framework as necessary for full implementation. The report on these country pilot studies was published by UNCTAD (2013). Three of the six key messages cited in the summary paper are:

1. *Private sector investment*, including domestic and foreign direct private investment, when operated in a responsible manner, can be a *key driver of economic development, job creation and inclusive growth.*
2. Attracting and generating private sector investment requires that the general framework conditions, such as *a sound legal and regulatory framework for domestic and foreign investment*, and efficient associated procedures be in place.
3. Once such a sound framework is in place, developing country governments, working in collaboration with the private sector itself and with development partners, can maximize economic value-added and job creation from private sector investment by *establishing priorities and focusing scarce resources on specific industries, value chains or segments of value chains*. The indicator framework developed by the G20 High-Level Development Working Group aims to inform this process.

## 5.6 ADDRESSING NUTRITION SECURITY THROUGH BIOFORTIFICATION

*Qualitative* assessments of crop varieties released across the world have taken a poor second place to *quantitative* characteristics (grain yield in particular, for instance IR8 rice). The search for the higher yield has been responsible for successfully addressing food insecurity in many countries and is to be applauded. Yet an equal effort is needed to select "biofortified" varieties on the basis of their relative nutritional value, and to endeavor to combine high yield potential with high nutritive value in the same genotype.

Bioenrichment of crops is potentially a sustainable means of improving the diet of a population, directly enhancing the resilience of a population to succumbing to undernutrition. It involves selection for elite nutritional traits among existing adapted "varieties" within a crop in a particular agroecological zone, or alternatively the more time-consuming creation of new "varieties" through hybridization or allied techniques. The biochemical composition of these selected varieties would rank highly in a nutritional factor which in the "normal diet" is in short supply and limits normal growth and development—in this case in humans. The instance of "Golden Rice", rich in pro-Vitamin A, was mentioned in Section 4.7.3.

### 5.6.1 Northern Nigeria, as a Country Case

There are many agricultural research centers in Nigeria, West Africa and elsewhere which can source genetic material for testing, under northern Nigerian conditions. There is also huge scope for advocacy on this issue, through mobilizing a group of breeders who understand the technical issues concerned and anthropologists who can deal with the cultural determinants of demand and acceptability among the population.

For instance, the limiting nutritional factor in which a variety is rich could be one of the "limiting" amino acid "building blocks" which our body cannot itself make yet needs to synthesise human proteins, or the precursor of Vitamin A (which the body can use to make Vitamin A), or thirdly it could rank highly in content of a mineral such as iron, which it absorbs from the soil in greater amounts than does the "control variety". Such high values have a genetic basis, such that this sought-after feature can be passed from one crop generation to the next. Of course, the "genotype" of these varieties which may be high in such variables, also has to be one which satisfies the requirement of being adapted to the environment in which it grows—it needs to compete well with the other good features of the "control" variety(ies) which a farmer is already growing, in terms of yield, tolerance to field pest or disease, and postharvest characteristics such as resistance to pests in the granary and cooking quality, and complies with the social norms of the community such as grain shape or color.

Some examples of bio-enriched varieties of crops which have found acceptability in parts of Nigeria, for instance, are high-lysine corn[37], high-beta-carotene (a precursor of Vitamin A) orange- and yellow-fleshed sweet potato, and high provitamin A cassava. The latter is being promoted currently in Nigeria, with funds from the Gates Foundation and the involvement of a private group called Harvest Plus.[38] Since 2011, there have been 4 pilots in Oyo, Imo, Benue and AkwaIbom States, and plans were afoot to extend to 8 states by the end of 2014. Harvest Plus is collaborating with the Ministries of Health and Agriculture in this venture, the effort much appreciated by the international nutrition organization GAIN. The distribution of this cassava

---

[37] Lysine is often a "limiting" amino acid in the human diet.

[38] www.harvestplus.org.

variety is linked to a sophisticated awareness campaign, including videos made by Nollywood actors, these videos having been released in late September 2014 (this modality seemingly more effective than posters and handout fliers).

These three examples of biofortified crops are indeed good news for Nigeria as a whole. Yet the cassava crop and its many derived prepared foods (such as *garri* and *fufu*), are relatively uncommon in the northern States. When it is present, it is fenced with a clay bank and brushwood to keep the animals from browsing its leaves.[39] The crop may cover less than 1% of the cropped area outside of the Lake Chad floodplain. However, as the cassava is all locally consumed at household level, compared with much of the cereal grain which is exported to the south via Katsina and Kano cities, it would be useful to encourage Harvest Plus to target northern States also as part of its high pro-Vitamin A cassava initiative, and facilitating State agricultural authorities to create a demand.

Sweet potatoes are similarly uncommon in northern Nigeria, perhaps also related to their leafy haulms and sweet-tasting tubers being a favorite of livestock, which during their free range grazing and browsing would soon put an end to the crop. Hence, the promotion of the orange-fleshed variety would be inappropriate to boost Vitamin A content of the diet for the majority in rural northern Nigeria, despite its undoubted dietary qualities (Ukpabi et al., 2012). Biofortified high-lysine corn offers a somewhat better option for north Nigeria, yet corn tends to be eaten more like a snack in the boiled or roast form, and is nowhere near as popular as sorghum or millet.

It is the latter two crops which would most warrant attention for biofortification to improve the status of limiting nutrients in the northern Nigerian diet. Yet, while there are high-lysine sorghum varieties grown in Ethiopia, the current author has found no evidence that these have been tested in Nigeria, or used in a breeding and/or selection program there. Such would surely repay applied research—selection based on grain yield or combined grain and straw yield is insufficient. Identification and popularization of adapted high-lysine sorghums would likely do much to promote good nutrition of all mothers and children across northern Nigeria and beyond, a nutrition-sensitive agricultural intervention which would build resilience against nutrition insecurity.

## 5.7 FOOD SAFETY

### 5.7.1 Introduction

Food safety refers to a number of related issues which determine whether the food we eat is free from contaminants or organisms which could lead to

[39] The abundance of semi free-range livestock across northern Nigeria, compared with the south, is manifest of a comparative advantage livelihood choice made by farmers over many years.

bodily infections. This can happen at any point along the food chain—crop or livestock production, food processing and storage, preparation for the "table" and handling. The terms food *safety* and food *quality* are sometimes confused. *Food safety* refers to all those hazards, whether chronic or acute, that may make food injurious to the health of the consumer, and is non-negotiable. Foods should not contain chemical, biological or physical hazards which can cause adverse effects on human health. By contrast, *food quality* includes all other attributes that influence a product's value to the consumer—negative characteristics such as spoilage, contamination with soil, discoloration and off-odors, and attributes which can be positive such as origin, color, flavor and texture.

Concerning *chemical* hazards, at field level, if good quality recommended agrochemicals are over-used, those absorbed by the crop from a drenching of insecticide, say, will be absorbed at levels which are dangerous to humans when the harvestable parts are ingested. Similarly with livestock, if vaccines are used above the recommended doses, humans can be poisoned through eating the meat, milk or eggs, and consuming fish which have themselves been poisoned sub-lethally by pollution in waterways. In countries with weak food regulations and enforcement regimes, the chemicals themselves can be adulterated and not comply with the description on the label (if a label is affixed at all), sometimes too dilute to be of any use while at other times, highly toxic.

With regard to *biological* hazard, storage of primary product food in poor quality warehousing, can lead to the grain becoming damp because of a leaking roof, for example. Groundnuts when damp are prone to rotting by a fungus called *Aspergillus* (*Aspergillus flavus* and *Aspergillus parasiticus*), which can lead to a carcinogenic toxin called aflotoxin to develop, this first understood in the United Kingdom in 1960 when 100,000 turkeys fed on groundnut meal died. Food products which can be contaminated with aflatoxins include cereals (corn, sorghum, pearl millet, rice and wheat), oilseeds (groundnut, soybean, sunflower, walnut and cotton) and milk.

During preparation of food in the manufacturing unit or kitchen, food exposed to flies or insecticide used to kill flies, can lead to intestinal afflictions or poisoning. Poor hygienic practices at home can also lead to diarrhea of infants and undernutrition, with contaminated drinking water often being the culprit. Had the water been boiled or filtered first this could have been avoided. Often too, national grid unavailability or outages, or fuel supply constraints or prices, do not allow for adequate refrigeration.

Concerning *physical* hazards, one example that is fresh in the memory of this author is the collection of nails and other metallic objects that were trapped by a bar magnet during initial conveyor belt screening of leaves from a tea plantation at a processing factory in Kampala, to which he had taken his students for a practical lesson. The tea company had no idea as to the origin of these unwanted artefacts.

The five key principles of food hygiene, according to WHO are as follows:

1. Prevent contaminating food with pathogens spreading from people, pets and pests (flies, rats, cockroaches etc.).
2. Separate raw and cooked foods to prevent contaminating the cooked foods.
3. Cook foods for the appropriate length of time and at the appropriate temperature to kill pathogens.
4. Store food at the proper temperature.
5. Do use safe water and raw materials.

Street food is particularly prone to contamination, often with minimal regulation enforced by a government standards agency. Such food is generally prepared and sold under unhygienic conditions, with limited access to safe water, sanitary services or garbage disposal facilities, and poses a high risk of food poisoning due to microbial contamination, as well as improper use of food additives, adulteration and environmental contamination. And even cottage industry processing premises are often ill-equipped and -regulated to maintain food safety and quality in a scientific and sustained manner.

In some countries, official monitoring has not worked well, as the risk of being caught and punished severely is outweighed by the potential commercial gains to be made by flouting the law. The rampant corruption which often exists offers a cheaper option also, in which misdemeanors are made good by informal payments so they do not reach court, or when they do, the punishment is reduced.

China is one such country. In 2008, six babies died and 300,000 were left very ill after consuming milk and infant formula milk products contaminated with melamine, an industrial chemical used in fertilizers and plastics. Because it is rich in nitrogen, melamine can be used by disreputable manufacturers to disguise milk that has been watered-down, such that tests for protein levels are falsely convincing. Consumed by humans, melamine can cause kidney stones and other potentially fatal conditions, especially in children. The scandal caused outrage in China and undermined public confidence in the local food industry, and in the government's ability to regulate it. In late 2008, the World Health Organization blamed the melamine scandal on fundamental failings among agencies responsible for monitoring food safety, including poor staffing and a lack of resources. China's premier Wen Jiabao vowed to prevent anything like it happening again, yet scares over contaminated foods remain regular occurrences, with the dairy industry in particular a frequent culprit.

In June 2012, for example, China's Yili group recalled batches of infant formula found to contain what it described as "unusual" levels of highly toxic mercury. While some of these scares might plausibly be categorized as accidental contamination, the melamine scandal as well as many cases before and since were deliberate breaches in pursuit of profit. Indeed, food and other product safety scares have become so prevalent in China that in 2012 at least two independent smartphone apps were launched to help Chinese consumers keep track of which products to avoid (Lim, 2013). Parents in mainland

China now favor imported baby formula in hope of better reliability, resulting in a scarce supply of this in Hong Kong.

Yet another criminal practice which has come to light in recent years in China concerns what has become known as "sewer oil". Following his undercover investigation in March 2010, Food Science Professor He Dongping of Wuhan Polytechnic University estimated that one in ten of all meals in China were cooked using recycled oil, often salvaged from the drains beneath restaurants. Though the State Food and Drug Administration issued a nationwide emergency investigation into the scandal, public confidence in the local food industry was further undone.

As Lim (2013) *ibid.* points out, the tainted milk scandal of 2008 had its origins in the rapid and largely ramshackle growth of the Chinese dairy industry, and laid bare some of harsher realities behind the country's economic miracle—exposing the high price that can be exacted for entrenched, institutionalized corruption and inadequate oversight during times of rapid growth.

### 5.7.2 Ways of Countering Malpractice

Such dangers as itemized above can be combatted in a number of ways. Minimalist use of agrochemicals during IPM regimes is popular now, and in the highest form of "organic agriculture", the use of all but a few specialist products would disqualify the farmer from eligibility for an "organic" certificate for the resulting produce, even inorganic NPK fertilizers. Plant breeders have also made progress in developing cultivars of crops which are more tolerant of pests and disease so that yield decrease is modest even if pesticides or fungicides are used sparingly or not at all.

To improve hygienic practices in the home, and the importance of education and creating awareness of those preparing food in poor communities (almost exclusively women), has been mentioned in Section 4.7.

Having a good food control system confers two main advantages to a developing country—to boost the health of its own citizens, and also to increase its revenue from the international food trade, by complying with international standards for food safety (together with branding, packaging and labeling). However, in emerging national economies, food safety is not a priority policy, often due to the shortage of economic and technical resources needed. This is hardly surprising, as the US Federal Food and Drug Administration, despite having a budget for FY2015 of $4.7 billion, can hardly cope with the large influx of food imports from abroad.

A food control system needs to be run by a competent well-resourced standards agency, which can both test agrochemicals (for strength or unwarranted additives) and produce (for pesticide residues, say) on the market and scrutinize labels, and for there to be an empowerment agency to ensure that those who carelessly or deliberately flout regulations are brought to book. The Food Safety and Standards Authority of India, established under the

Food Safety and Standards Act, 2006, is the regulating body related to food safety and laying down of standards of food in that country, for instance.

Food control systems should cover all food produced, processed and marketed (including that imported) within a country. The control system will typically comprise (1) a food law and regulations, which are enforceable. The law should contain legal definitions of unsafe food, and the prescription of enforcement tools for removing unsafe food from commerce and punishing responsible parties. *Also, governments need food standards and a clear mandate and authority to prevent food safety problems developing* (2) policy and operational coordination at national level (3) inspection services (4) laboratory services for food monitoring, and (5) related information, education, communication and training capability.

The foremost responsibility of food control is to enforce the food law(s) protecting the consumer against unsafe, impure and fraudulently presented food, by prohibiting the sale of food not of the quality demanded by the purchaser. Consumers expect protection from hazards occurring along the entire food chain, from primary producer through consumer (often described as the farm-to-table continuum). Protection will occur only if all sectors in the chain operate in an integrated way, and food control systems address all stages of this chain.

Responsibility for food control in most countries is shared among different agencies or ministries. The roles and responsibilities of these agencies may be quite different and duplication of regulatory activity, fragmented surveillance and a lack of coordination are common. There may also be wide variations in expertise and resources across agencies, and the responsibility for protecting public health may conflict with obligations to facilitate trade or develop an industry or sector.

FAO and WHO are the two main specialized agencies of the United Nations involved in food safety and quality technical cooperation programs with developing countries.[40] FAO assistance in food control and food standards is a major activity and is delivered at global, regional and country levels. Published manuals cover a range of different aspects of food control systems, and meetings, seminars and workshops are conducted in all regions of Africa, Asia and the Pacific, LAC, Eastern Europe, the 'Near East' and North Africa. Technical assistance is provided in many areas such as the following:

- establishing or strengthening national food control systems and infrastructure;
- assistance in preparation of food law and regulations;
- workshops on developing national strategies for food control;
- assistance in establishing or improving food analysis capabilities;
- providing training in food inspection, analysis and food handling;
- providing training in management of food control systems.

[40] Retrieved from www.fao.org/docrep/006/y8705e/y8705e06.htm (accessed June 3, 2015).

WHO provides technical assistance at international, regional, and country levels. Under its decentralized structure, WHO is divided into six regions, with Regional Offices responsible for providing assistance to Member States to build capacity in developing and strengthening their National Food Safety Programs to safeguard consumer health. Such assistance includes the following:

- developing regional and national food safety policy and strategies;
- preparation of food legislation, food regulations and standards, and codes of hygienic practice;
- implementation of food inspection programs;
- promoting methods and technologies designed to prevent food-borne diseases;
- developing or enhancing food analysis capability;
- development and delivery of hygiene training and education programs;
- establishing healthy markets and enhancing the safety of street food; and
- promoting the establishment of food-borne disease surveillance activity.

Technical assistance in the food control area may also be obtained through the World Bank, other development banks and from bilateral donor agencies. Access to such funds is dependent upon the priority that developing countries attach to strengthening their food control systems as reflected in their national development plans.

## 5.8 THE DIFFICULTY OF ENHANCING THE STABILITY ASPECT OF FOOD SECURITY

Of the four dimensions of food security (availability, access, utilization and stability), the one in which least progress toward improvement has been made is "stability", reflecting the effects of growing political instability and international food price volatility. The contours of "stability" can also be subject to sudden change at the level of individual homes and livelihoods, with individuals unable to affect the cause, and faced with having to choose the least bad option as the favoured response.

As considered in Chapter 3 "Causes of Food Insecurity", the factors causing reduced food *availability* include natural- and man-made disasters, and climate change (discussed in more detail in Section 6.9). To these can be added factors which reduce the *access* component of food security, especially the unaffordability of basic commodities (as discussed in more detail in Case Study 9 on the book's companion website). And a people on the move, running from conflict, without their granary, cooking utensils and supply of safe water, cannot adequately attend to the *utilization* component of food security. As the causative factors mentioned are often unpredictable, they can bring *instability* to food security status. Nations and families have to cope as best they can with this instability, through ingenuity and their prescience in trying to manage risks.

As the FAO/IFAD/WFP *State of Food Insecurity in the World,* 2013 puts it "The *stability* dimension is divided into two groups. The first group covers factors that measure exposure to food security risk with a diverse set of indicators such as the cereal dependency ratio, the area under irrigation and the value of staple food imports as a percentage of total merchandise exports. The second group focuses on the incidence of shocks such as domestic food price volatility, fluctuations in domestic food supply and political instability".

The "fluctuations in domestic food supply" just mentioned, is often seasonal for rural populations, relating to the "hungry season", which occurs before the staple crop in the field is ready for harvest yet the homestead granary is empty or almost so. Such families are highly vulnerable to hunger after a poor season, due to drought or field pests like locusts or weaver birds, for example (the PSNP in Ethiopia adjusts its safety net contributions to address hungry season needs—see Case Study 1, para 1.2.3 on the book's companion website).

Another example of seasonal fluctuating domestic food supply, is that of shepherding families in the Nepalese Himalayas, who graze their animals above 3,000m (up to 5,000m in summer), and drive the sheep down to middle hill markets perhaps only once a year. This can provide a spectacular sight in Pokhara, western Nepal, for instance, just before the auspicious 2-week festival of *Dashain*, late September–early October. Hundreds of colorfully attired shepherds together with thousands of plump sheep jostle in the narrow streets of the town, where the animals are bought by urban residents and livestock traders. Each shepherd's liquidity peaks just this one time a year, and as the calendar year progresses, ever less money is available to buy crop produce from middle hill markets. These crops cannot be grown on the poor soils at 3,000m for more than 3 months a year, because of the low temperature. The nutrition country profiles produced by FAO and WFP indicate that among mountain and middle hill dwellers in western Nepal, the prevalence of stunting in children under 5 years of age is at least 50%, more pronounced than in the lowland *tarai*. Severity of child undernutrition may well fluctuate in sympathy with this pattern of liquidity, with the trough corresponding with the northern hemisphere summer and early autumn months before both the rice harvest and sheep sales, though data are not available to confirm this.

And consider this ... during the monsoon, as river levels rise in Bangladesh, a woman and her children climb onto the tin roof of their home-on-stilts on their tiny "char" island plot, well-aware they may get swept to their deaths. The island has already become submerged, and perhaps even the roof will be, but they are also aware that unless they stay, someone else may occupy that riverine island once the waters recede, and the precious silty island which provides sustenance for the family will be lost to others for ever, and they perish through lack of an alternative livelihood. The woman and her children take what they hope will be the lower risk, and stay ...

There are non-cyclic micro-level events too which affect stability of food and nutrition security, based on family circumstance, in which an "event" may impact one individual or family more than another. The death of an old widow's cow in rural Asia is a catastrophic event for that woman, whose days may be numbered from that moment. Suddenly, there is no milk in her diet or source of income with which to buy salt or cooking oil. Not least, the constant companion in her life is gone. Perhaps someone will take pity on her, perhaps not.

Yet another life-threatening example of a one-off event is sudden illness or death of a "bread-winner" or carer in a family, or an able-bodied relative moving to another country to seek employment in the Persian Gulf, say, so no longer available for family labor and the prospect of compensatory remittances "uncertain", until they start flowing. Farm productivity and production declines, and children receive less attention than before.

Families in Yola, the capital of Adamawa State in northern Nigeria in 2014 generously welcomed internally displaced people (IDPs) from Borno State in the northeast of the country who were fleeing south from the Boko Haram insurgency there. Kinship ties offered support, yet the host families were already stretched providing for their own food security. Basic safety net systems are in place, orchestrated through the American University of Nigeria campus in Yola, in tandem with State authorities, Church and Islamic foundations, yet the extra stress on household food security coping mechanisms will continue until Boko Haram is driven out of Borno[41], and the IDPs can safely return. Only then may "stability" of food security return to both host families and the IDPs.

There is a geographical component to food security stability too. West Asia, North Africa and the Caribbean are particularly vulnerable because of their limited or fragile natural resource base, and their heavy reliance on international food markets for domestic supplies. An example of this from Central America is the impact of Hurricane Mitch in late October/early November 1998 on regional markets and the stability of food supplies. Honduras, Nicaragua and Guatemala were particularly badly hit. At least 11,000 people died, making Category 5 Mitch the most deadly storm in the Western Hemisphere since the Great Hurricane of 1780 in the eastern Caribbean. Additionally, several million people were made homeless or severely impacted. The hurricane's largest direct impact was the destruction of transport infrastructure—27,754 km of roads and rail tracks, and 156 bridges. As a result, during the period immediately after the hurricane, flows of food and other cargo to those markets were interrupted, and the cost of urgent items rose temporarily.

At the time of Hurricane Mitch, trade outside of the region was not regarded as a priority by the countries concerned. With the exception of Nicaragua, nor did the countries affected by the hurricane maintain food

[41] Whilst reading the proofs of this book, the author received a plea from his friend in Maiduguri, capital of Borno State. His home in town had become over-run with relatives and friends fleeing from Boko Haram gangs in the rural areas that he had no food left to feed either his family or IDPs, so could I provide help urgently?

balance records that might have provided an indication of the status of national inventories and the rate of consumption relative to the availability of basic foods. As a result, it was not possible to gain a clear sense of how to formulate a national import policy or a policy for the promotion of domestic production, or even to gain some idea of how vulnerable the population may be to cyclonic events. In such a situation, when disaster strikes, any planning of food imports or donations has to be based on an emergency basis rather than being strategic, without there being any surety of available domestic capacity.[42]

## REFERENCES

Acord, 2014. Why wait until the next food crisis?—Improving food reserves strategies in East Africa (Curtis, M.), 60 pp. <www.acordinternational.org/silo/files/why-wait-until-the-next-food-crisis-.pdf>.

ActionAid, 2011. No More Food Crises: The Indispensable Role of Food Reserves. ActionAid International Briefing, ActionAid UK, London, 17 pp.

ActionAid USA, 2013. Rising to the challenge: changing course to feed the world in 2050 (Wise, T.A., Sundell, K.), 28 pp. <http://www.actionaidusa.org/publications/feeding-world-2050>.

Ashley, J.M., 1999. Food Crops and Drought. The Tropical Agriculturalist Series. Macmillans, UK, 133 pp.

Ashley, J.M., Jayousi, N., 2006. Setting a Palestinian National Food Security Strategy. Palestine-Israel J. 13 (3), 112–118.

Ashley, J.M., Khatiwada, B.P., 1992. Local wheat makes the grade in Nepal. Appro. Technol. 18 (4), 30–32.

Ashley, J.M., Shugaba, M.A., 1994. Coping with drought in north-eastern Nigeria. Appro. Technol. 21 (2), 30–33.

BBC News, 2015. Turning Ethiopia's desert green. (Haslam, C.), April 20, 2015. <http://www.bbc.co.uk/news/magazine-32348749>.

Beman, J.M., Arrigo, K.R., Matson, P.A., 2005. Agricultural runoff fuels large phytoplankton blooms in vulnerable areas of the ocean. Nature. 434 (7030), 211–214.

Borlaug, N., 2007. Feeding a hungry world. Science. 318 (5849), 359.

Collaborative Partnership on Forests, 2014. Independent Assessment of the International Arrangement on Forests: Report of the Team of Independent Consultants. September, 2014, 172 pp.

De Schutter, O., 2013. Bali package must allow ambitious food security policies. UN Expert on WTO Summit. Press Release, December 2, 2013, Geneva. <www.srfood.org/en/bali-package-must-allow-ambitious-food-security-policies-un-expert-on-wto-summit>.

Donald, C.M., 1968. The breeding of crop ideotypes. Euphytica. 17, 385–403.

Evenson, R.E., 2003. The green revolution in developing countries; an economist's assessment, 39 pp. <http://unpan1.un.org/intradoc/groups/public/documents/APCITY/UNPAN020402.pdf>.

FAO, 1997. Strategic grain reserves—guidelines for their establishment, management and operation. FAO Agric. Serv. Bull. 126, Rome.

FAO, 2009a. Responding to the Food Crisis: Synthesis of Medium-Term Measures Proposed in Inter-Agency Assessments. FAO, Rome.

FAO, 2009b. Country responses to the food security crisis: nature and preliminary implications of the policies pursued. Initiative on Soaring Food Prices. FAO, Rome.

FAO, 2011. Global Food Losses and Food Waste: Extent, Causes, and Prevention, Rome.

[42] Retrieved from www.fao.org/docrep/003/y2784e/y2784e05.htm (accessed June 5, 2015).

FAO, 2014. CFS Principles for Responsible Investment in Agriculture and Food Systems First Draft (For Negotiation), 13 pp.

FAO, IFAD and WFP, 2013. The State of Food Insecurity in the World 2013: the multiple dimensions of food security. Executive Summary.

Finegold, C., 2009. The importance of fisheries and aquaculture to development. In: KSLA (2009). Fisheries, Sustainability and Development. The Royal Swedish Academy of Agriculture and Forestry. pp.353–364. <www.ksla.se/wp-content/uploads/2011/10/Fisheries_sustainability_and_development.pdf>.

Fritz, T., 2014. Putting Food Security Before Trade: The WTO and the Conflict Over Food Reserves. Forschungs- und Dokumentationszentrum Chile-Lateinamerika (FDCL), Berlin, April 2014, 16 pp.

Gilbert, C.L., 2011. Food Reserves in Developing Countries. International Center for Trade and Sustainable Development, Geneva, Issue Paper No. 37, 41 pp.

Government of Rwanda, 2011. National Post-Harvest Staple Crop Strategy. Ministry of Agriculture and Animal Resources, 56 pp.

HLPE, 2012. Social protection for food security. A Report by the High Level Panel of Experts on Food Security and Nutrition of the Committee on World Food Security, Rome 2012. HLPE Report 4, June 2012. 100 pp. Retrieved from: <www.fao.org/fileadmin/user_upload/hlpe/hlpe_documents/HLPE_Reports/HLPE-Report-4-Social_protection_for_food_security-June_2012.pdf> (accessed 28.05.15.).

IFPRI, 2006. Grain marketing parastatals in Asia: results from six case studies. World Dev. 35 (11), 1872–1888.

IFPRI, 2011. Strategic Grain Reserves in Ethiopia, Institutional Design and Operational Performance. IFPRI, Washington (http://www.ifpri.org/publication/strategic-grain-reserves-ethiopia).

ISAAA, 2013. Brief 46-2013, Executive Summary, Global Status of Commercialized Biotech/GM Crops: 2013 International Service for the Acquisition of Agri-Biotech Applications (accessed 5.06.15.).

Lim, V., 2013. Tainted Milk: Unravelling China's Melamine Scandal. Think Business website, National University of Singapore. Retrieved from <http://thinkbusiness.nus.edu/articles/item/118-tainted-milk-unravelling-china5E2%80%99s-melamine-scandal> (accessed 01.06.15.).

Mousseau, F., 2010. The High Food Price Challenge: A Review of Responses to Combat Hunger. The Oakland Institute, Oakland, CA (http://www.oaklandinstitute.org/pdfs/high_food_prices_web_final.pdf).

Murphy, S., 2009. 'Strategic Grain Reserves in an Era of Volatility'. Institute for Agriculture and Trade Policy, Minneapolis, MN, 16 pp.

Schaffnit-Chatterjee, C., 2014. Agricultural Value Chains in Sub-Saharan Africa: From a Development Challenge to a Business Opportunity. Current Issues—Emerging Markets. Deutsche Bank Research, 28 pp.

Ukpabi, U.J., Ekeledo, E.N., Ezigbo, V.U., 2012. Potential use of roots of orange-fleshed sweet potato genotypes in the production of B-Carotene rich chips in Nigeria. Afr. J. Food Sci. 6 (2), 29–33.

UNCTAD, 2011. The Potential Establishment of Emergency Food Reserve Funds (Lines, T.), December 2011, 26 pp.

UNCTAD, 2013. Indicators for measuring and maximizing economic value added and job creation arising from private sector investment in value chains: summary of pilot results and implications for indicator methodology. Final Report to the High-Level Development Working Group. May 2013, 16 pp.

UN-Habitat, 2014. The State of African Cities 2014: re-imagining sustainable urban transitions.

World Bank, 2012. The grain chain: food security and managing wheat imports in Arab countries. Working Paper, disclosed April 16, 2012. World Bank and FAO. 80 pp.

Chapter | Six

# Cross-Cutting Issues

## 6.1 RIGHT TO FOOD

The right to adequate food is defined by General Comment number 12 of the UN Committee on Economic, Social and Cultural Rights, in its 20th session in 1999. The Committee declared in a 41-point declaration that "the right to adequate food is realized when every man, woman and child, alone or in community with others, has physical and economic access at all times to adequate food or means for its procurement". Obligations of States (and hence their citizens), the UN and other International Organizations are listed.

The 41st session of the Committee on Food Security (CFS) in October 2014, entitled "Right To Food—10-year Perspective", provided the opportunity for some Member Countries (El Salvador, India and Jordan) to share their national experiences in implementing the Right to Food Guidelines of 2004, and for all CFS Member Countries to reaffirm their commitment to implement those Guidelines and strive for the realization of the right to adequate food of all in the years to come. A seven-point memorandum was issued by the Committee (see Box 6.1).

FIAN International (Food First Information and Action Network)[1] has conducted country studies in Guatemala, Nicaragua and Philippines *inter alia*, on national policies relating to the right to food. Brazil has been a leader in implementing the right to food, with its Federal Law for Food and Nutrition Security of 2006, and its cornerstone *Bolsa Familia* Program, which benefits 11 million poor families. In return for a cash transfer, which allows for food purchases and some immediate relief from the difficulties created by poverty, families must keep their children in school, giving them the opportunity to break the inter-generational poverty cycle.[2] The FAO Right to Food webpage provides useful information on right to food developments and FAO efforts to promote them.[3]

---

[1] Retrieved from www.fian.org (accessed June 5, 2015).
[2] Retrieved from www.worldhunger.org/articles/08/hrf/ananias.htm (accessed June 5, 2015).
[3] www.fao.org/righttofood/index_en.htm.



Food Security in the Developing World. DOI: http://dx.doi.org/10.1016/B978-0-12-801594-0.00006-3

**Box 6.1 CFS Memorandum on the Right to Food (2014)**

1. Welcomes the significant contribution of the Voluntary Guidelines for the Progressive Realization of the Right to Adequate Food in the Context of National Food Security in guiding national governments in the design and implementation of food security and nutrition policies, programs and legal frameworks in the last ten years, and reaffirms its commitment towards achieving the progressive realization of the Right to Food in the years to come.
2. Encourages all CFS stakeholders to promote policy coherence in line with the Voluntary Guidelines for the Progressive Realization of the Right to Adequate Food in the Context of National Food Security, and in that context, reaffirms the importance of nutrition as an essential element of food security;
3. Reaffirms the importance of respecting, protecting, promoting and facilitating human rights when developing and implementing policies and programs related to food security and nutrition;
4. Invites to consider a human rights based approach to food security and nutrition and encourages to strengthen mechanisms that facilitate informed, participatory and transparent decision making in food security and nutrition policy processes, including effective monitoring and accountability;
5. Urges all CFS stakeholders to afford the highest priority to the most vulnerable, food insecure and malnourished people and groups when designing and implementing food security and nutrition policies and programs;
6. Urges all CFS stakeholders to integrate gender equality and women's empowerment in the design and implementation of food security and nutrition policies and programs;
7. Underscores the important contribution of non-government stakeholders in the design, implementation, monitoring and evaluation of food security and nutrition policies and programs at all levels.

*Source:* Memorandum from 41st session of the Committee on Food Security (CFS) in October 2014, entitled "Right To Food – 10-year Perspective". Reproduced with permission from FAO.

IDS (2014) points out that despite an awareness of the right to food as a vital foundation concept for all food security thinking and interventions, there is a persistent failure to translate this commitment and understanding into concrete actions. Likewise, the rights enshrined in the Convention on the Elimination of all Forms of Discrimination Against Women (CEDAW), the Universal Declaration of Human Rights and other relevant agreements are not being respected at the level of implementation in the context of food and nutrition security programs. Unless the challenge of hunger and undernutrition is framed in a more comprehensive way that captures these complex gendered dynamics and is grounded in human rights, IDS opines that food security and nutrition interventions will do little to challenge the fundamental imbalance that has generated a situation of food scarcity for many, in a world that produces more than enough nutritious food for everyone. They can do little to address the specific gender dimensions of food insecurity, or transform gender inequalities that both perpetuate and result from hunger and undernutrition (see also Section 6.8, relating to food sovereignty).

# 6.2 GENDERING FOOD SECURITY

## 6.2.1 Introduction

Women have several vital roles in building resilience against food insecurity crises in the developing world, at household and community levels. In rural areas, home to the majority of the world's hungry, as well as taking care of the home and looking after children, they grow most of the crops for domestic consumption and are primarily responsible for all postharvest activities, including preparing and processing food. They also handle livestock, gather fodder for stall-feeding and fuelwood, and manage the domestic water supply. Moreover, they undertake a range of community-level activities that support agricultural development, such as soil and water conservation. Yet women's work often goes unrecognized by their menfolk at home, and in the public domain they lack the leverage necessary to gain access to resources, training and finance.

Female-headed households are increasing in number globally, due to civil conflict and natural disaster, disease or death of the male partner or his migration from rural areas in search of work. The Asian Development Bank (2013) highlights that this huge burden on, and achievement by, women is despite the many challenges which women and girls face, often embedded in society's norms and set in law. In many developing countries, women and girls are negatively affected through two main instruments, discrimination at home and in the community.

## 6.2.2 Discrimination at Home

Respect by society and family for girls, and expectations of them, are far lower than the equivalent accorded to boys. This leads to reduced access to education and employment opportunities, which curtails women's economic autonomy, weakening their bargaining position within the family. This is manifested in their having little or no voice in key household decisions, differential feeding and caregiving practices favoring boys and men, and lower health, nutrition and pregnancy outcomes. Family attitude that money spent on educating girls is "wasted" has an existential outcome, with studies from Africa showing that children of mothers who have spent at least 5 years in primary education are 40% more likely to live beyond the age of 5 years.[4]

An example from the current author's experience .... when working in the hills of western Nepal in the mid-1990s, he was involved with helping farmers select improved "varieties" of corn, wheat and rice via PVS trials on their own farms, using seeds from a basket of new entries sourced from research stations or smallholdings in another part of the low Himalayan hills. When scoring the new entry performance against the "control" landrace(s) which the family was using, it was noted how different were the value

[4] Retrieved from http://www.fao.org/gender/gender-home/gender-programme/gender-food/en/ (accessed June 10, 2015).

judgements applied by men and women of the farming households. The men were most concerned about *field* characteristics—what the crop looked like, its height, grain color and so on. By contrast, women were concerned with *postharvest* characters, such as how resistant the grain was to pest attack in storage, how long it took to cook (on which would depend how much firewood the women and girls needed to collect), the taste of the grain and the ease with which it dispelled hunger when eaten. In short, a far more practical take on "quality"! A characteristic on which both men and women usually agreed, was that any new entry should yield not less than the "control".

How essential then to seek the views of both men and women, to ensure that variety selection is made on a sound sustainable basis (initially the men saw no value in bothering to consult the women on their views ... while the women scoffed at the men's superficial grasp of the concept of crop improvement). The questions asked of farmers may need to be given to individual farmers, or gender-separated groups—mixed groups can sometimes be dominated by male opinions. Living in a rustic house in Sandikharka village the author was woken at 4:00 am each morning by the sound of those under-rated women in neighboring traditional houses stone-grinding grain into flour for the day's unleavened bread, the same women who would have been last to bed the evening before.

### 6.2.3 Discrimination in the Community

The second instrument of discrimination makes it more difficult for women to fulfill their vital roles in food production, preparation, processing, distribution and marketing activities. In agricultural practice, there are several gender-specific constraints that limit the ownership by women of productive assets (such as land), or access to inputs (such as fertilizers) and services (such as credit and agricultural extension). Some laws, such as those governing access to land, include inequitable and exclusionary provisions, thereby institutionalizing discrimination. According to a study by the World Bank (2010), 103 of the 141 countries studied have laws and cultural barriers that can thwart women's economic aspirations through limiting their access to key resources. These obstacles not only heighten their vulnerability to personal food insecurity but also reduce their contribution to family agricultural production.

Another example from the current author's experience—in western Uganda among the Batoro people around Fort Portal, agricultural land ownership is firmly under the control of men. Though women do more work than men in the fields, they do not choose the crops that are grown – almost exclusively corn, as a cash crop. Women find it difficult to grow the pulses and vegetables that they need to provide a balanced diet for the family, and because this is common, the price of each on the market is high, owing to shortage. In 1996 the author worked with some brave women who had agreed to club together, and devise a village women's cooperative. For this, they wrote a constitution which was accepted by a bank to secure a loan, with which they jointly bought a small patch of land on which they could grow the pulses and vegetables that their families needed. For this initiative, they expected to get beaten by their husbands when they found out, and they were. [By contrast, some educated Batoro women living in Fort Portal town are active as farm-gate buyers of corn from male farmers, which they then store in hired warehouses and transport by rail to Kampala when the grain price rises prior to the following harvest. The importance of education as a means of empowerment is clear, enabling those women to become "bread-winners" for their families].

Here is an example of discrimination which affects both the home and the community.... in northern Nigeria, the land is a fertile plain, supporting good crops, largely of millet and sorghum, with cowpea as the main pulse. However, much of the grain is not eaten in the far north but sold to the south of the country, northern Nigeria being known as the country's "breadbasket". This cash crop economy results in more than half of the children under 5 years of age in the north being stunted. This would not be the case, if cultural norms were different. Currently, women are not well-educated, and because of cultural norms many are not able to leave home to visit the market to buy the pulses and vegetables they need to feed their children, and men and boys are fed before girl children and mothers; the first child is not cared for by the mother but sent to a relative to look after, who does not prioritize its welfare, and standards of hygiene and sanitation are poor—diarrhea is considered the "norm", and so on. Awareness campaigns to educate the men to upgrade the value they give to women is much-needed, though not in demand by the men ! A senior member of the government in one northern State assured the author during his mission there in 2014 that he would never employ a woman in his department "Why should I when she'll get pregnant every year?!"—there was no acknowledgment of the role and attitude of her partner in insisting on many offspring, rather than encouraging child spacing.

## 6.2.4 Empowerment and Gender-Sensitive Policies

### 6.2.4.1 EMPOWERMENT ALONE

The examples above point to the perilous linkage between a community's agriculture and its food and nutritional security. In all instances, it was not only the lives of women which were affected but also those of the whole family and wider community, held in check by the patriarchal norms prevailing. Challenging the constraints that women face needs to be a key component in the fight against hunger and undernutrition. The cost to society (both its women and its men) of not acting decisively and urgently will continue to be considerable. However, *amendment of gender-discriminatory legislation is insufficient*. Social and cultural norms and the gendered division of roles they impose must be changed. *Empowerment* of women is required, to enable a greater role for women in decision-making at all levels, including the household, local communities and national Parliaments.

Women's empowerment is not only a priority goal in itself but a human right, already recognized as such in pledges and commitments by governments. Empowering women and girls is not just necessary for their well-being, but also a means to broader agricultural development and food security, and it is economically sound. Studies show that if women farmers were given the same access to resources (such as land, finance and technology) as are men, their agricultural yields could increase by 20–30%, national agricultural output could rise by 2.5–4%, and the number of undernourished people could be reduced by 12–17% (UNDP, 2012). Yet, is empowerment alone sufficient?

##### 6.2.4.2 GENDER TRANSFORMATION OF THE FOUR COMPONENTS OF FOOD SECURITY

Food security for all cannot be achieved until the four pillars of food *availability*, *access*, *utilization* and *stability* are adequately realized. Yet this framework is rarely sufficiently gender-aware to ensure that food security will in fact be universal (IDS, 2015). For example, the current policy preoccupation with increasing the *availability* of food is through the short-term strategy of providing food assistance, and longer-term strategy of boosting availability through greater productivity.

The policy focus on *availability*/agricultural production inadequately engages with the need to achieve good nutritional outcomes at the *individual and household* level. At the micro-level, policymakers and food security programs increasingly include women by investing in small-scale women's enterprises, for example. There is also a trend for better recognizing the need to increase women's access to productive resources, such as land, water and credit. Though these are welcome steps, they do not go far enough, and risk valuing women merely as an instrument to improve agribusiness efficiency. The micro-measures fail to recognize that *economic empowerment is only part of the solution to improve women's food security*. Unless mitigating measures also aggressively attack the root causes of gender inequality, there is a danger that current approaches will add yet more to women's current (often unpaid, always under-recognized and under-valued) workload. The cycle of gender discrimination would likely continue, perpetuating gender-injustice, poverty and food insecurity.

Tackling people's social and economic *access* to food and good nutrition also requires a more politically-engaged approach, which challenges head-on the gender dimensions of poverty, addressing gender-inequitable power relations and norms, inside the household and beyond. Social protection programs such as cash and voucher, and food-for-work schemes may be empowering for women in some respects but are often not *gender-transformative*, instead often reinforcing the existing *status quo* of unequal gender roles and relations. Other forms of support may be required for farming geared toward satisfaction of family and community food needs, such as serving the values of livelihood resilience, autonomy and stability rather than being oriented toward profit and market competiveness.

Improved income security for women should follow from better access for women to waged employment—on-farm and off-farm, in both the informal and the formal sectors. Women are more likely to spend their personal incomes on food and children's needs—research has shown that a child's chances of survival increase by 20% when the mother controls the household budget. Other studies have shown a direct correlation between increased incomes for women and improvements in household food security. IDS (2015) provides inspiring examples of participatory gender-transformative interventions in Gujarat, India (regarding empowering women with the right that the government issues food entitlement ration cards in the woman's name, rather than name of her male partner "head of household"), and in Latin America among Maya Chorti communities in the border municipalities of Guatemala, Honduras and El Salvador (with women empowered to grow crops of their choice in kitchen gardens).

Education and employment policies need to be more gender-sensitive, thereby removing one of the obstacles in the way of access to food for women. Girls' limited schooling reduces employment opportunities and increases their vulnerability to discriminatory practices. The only way to transform this vicious circle into a virtuous circle is to ensure better girls' education outcomes and equitable compensation for women workers. Yet policies alone may achieve little, without a supporting enabling environment, through far more investment in potable water supplies, for instance. Droughts caused by climate change are eliminating existing water supplies, making the distance from home to the water source even longer, and traditionally in developing countries collecting water for the family is the work of women and girls. In the developing world they travel on average 10–15 km to reach the water source, spending up to 8 h a day doing so, according to UNIFEM. As a result, millions of girls around the world are not sent to school, whatever the national enabling policy on education might be.

The gender aspect of the *utilization* pillar is also highly relevant, given that the majority of undernourished people in the world are women and girls. Not being hungry does not necessarily mean being nutritionally-secure, as discussed in Chapter 2 "Manifestations and Measurement of Food Insecurity". Women are often the victims of food discrimination. In many households and communities, women and girls eat only the food that is left after the males in the family have eaten, this frequently resulting in their undernutrition. In parts of South Asia, for instance, men and boys consume twice as many calories, even though women and girls do much of the heavy work. A study in India found that girls are four times more likely to suffer from acute undernutrition than are boys. The physiological needs of pregnant and lactating women also make them more susceptible to undernutrition and micronutrient deficiencies. Twice as many women suffer from undernutrition as men, and girls are twice as likely as boys to die from it. Maternal health is crucial for child survival—an undernourished mother is more likely to deliver an infant with low birth weight, significantly increasing its risk of dying, and even the mother herself dying in childbirth as her pelvis is small.

In accordance with the common gender division of labor within the homestead in rural areas of the developing world, women play an essential role in food processing (fruit drying, cheese and yoghurt, smoked fish, etc.). Processing fosters diversity of diet through the year. It often renders the food more digestible, minimizes food waste and losses in the home, adds value and improves marketability of primary produce enabling an optional income stream. Women also have a major role in food distribution (physical access), looking after post-harvest threshing and food storage on a day-to-day basis, and because they also prepare and cook the food have a predominant responsibility for *food safety*. Some of this workload surely needs to be shared with menfolk, as it very often is in developed countries, though this would require a seismic shift in male attitudes toward women. Efforts to improve food safety must take into account existing gender roles in the food chain. Training/coaching and mentoring women in hygiene and sanitation can make an immediate contribution to household and community health.

Education in general (of both women and men) can play a major role in improving the status of women, the nutrition of their families and national food production. *A cost-benefit analysis carried out by the World Bank indicates that investment in the education of females has the highest rate of return of any possible type of investment in developing nations.* It results in higher productivity, reduced fertility, reduced child morbidity and mortality rates, and increased application of environmental protection measures. *In the state of Kerala, India, a longstanding commitment to the education of females has been cited as a major factor in increasing life expectancy to 70 years, compared with the Indian average of 56–58 years.*[5] FAO's Gender and Development Plan of Action includes a number of commitments aimed both at improving women's access to adequate nutrition and at providing them with the knowledge and resources they need to improve their families' nutritional status.[6]

There are very many ways that can save the amount of time and effort that women in the developing world spend on menial labor-intensive tasks, which would enable them to devote more time to preparing nutritious meals at home, for instance, and even have some "free time" at which their menfolk are usually adept ! Several ways are mentioned in the book, such as selecting "improved" varieties of crops that cook more quickly requiring less fuelwood to be collected (see Section 6.2.2); conservation agriculture in southern Africa (see Case Study 2 on the book's companion website), and ox draft

[5] Retrieved from http://www.fao.org/docrep/x0262e/x0262e16.htm (accessed June 10, 2015).
[6] Retrieved from http://www.fao.org/docrep/005/y3969e/y3969e03.htm (accessed June 10, 2015).

plowing in Nepal, which can reduce the time women need to spend cultivating the land by hand hoe (see Section 7.5); and, the manual oil palm fruit presses in Liberia (see Section 4.5.3).

Other examples of "tools" that can be used are a simple hand-held corn sheller to strip grain off the cob so much more easily than using the fingers, and better still a mechanized sheller (see Fig. 6.1); a rocking arm groundnut sheller rather than this being done manually; "rediscovering" the old technology of the hydraulic ram pump, a simple device lowered into a moving flow of water that can shoot a jet of water up a hill through a narrow gauge pipe, which will save women carrying heavy water containers from the stream to a

**FIGURE 6.1** A mechanized sheller of corn cobs, easier and far quicker than hand shelling in Hoddaidah governorate Yemen. Bulking improved seed for distribution to farmers as planting material. Photos: Jafaar Hasan Alawi Al-Jeffri.

settlement on a hill which has no well; and, paddle wheels, that like the ram pump, using only the energy of flow in the stream, can lift water from stream-level into a channel on the river bank, for irrigation or domestic use. Facilitators (champions) at the village level will be needed to get action on these interventions, and linkage with a vocational college workshop or village blacksmith may help.

Furthermore, climate change and spreading conflict are compounding factors which impact the *stability* of food production, distribution and consumption, with often disproportionate impacts on women and girls. Women often act as "shock absorbers" in times of crisis, restricting their own food intake so that others can eat. Likely future impacts of climate change on women and children, in particular, need assessing, as already done for Yunnan Province in China, for instance, where the switch to cash crops and women migrating for work feature prominently (Opitz-Stapleton, 2014).

#### 6.2.4.3 INVESTING IN GENDER MAINSTREAMING

Explicitly missing from the standard FAO definition of food security and the "four pillars" are issues of gender inequality and the right to food, along with gender-based violence, women's unpaid care responsibilities, and the extra burdens thrust on carers (who are normally women) by HIV and AIDS. The standard apolitical definition of food and nutrition security (see Section 1.1) leads to insufficient and mis-targeted policy responses and a failure to both transform gender inequalities and realize the right to food for *all* people.

A huge compendium produced by the World Bank points to gender-based inequalities all along the food production chain "from farm to plate", which impede the attainment of food and nutritional security. Maximizing the impact of agricultural development on food security involves enhancing women's roles as agricultural producers as well as the primary caretakers of their families (World Bank, 2009).

IDS (2015) *op. cit.* points out that the gender equality potential of short-term solutions is being limited by narrow targeting of groups perceived as the most vulnerable, with an almost exclusive focus on pregnant women and very young children. As a result, the needs of other vulnerable groups, such as older women, adolescent girls, women from marginalized communities, single women (and men) and women with disabilities, are rendered invisible.

Even the longer-term strategies applied are short-sighted. A focus on increasing agricultural production is certainly needed, but at present there is a failure to fully understand and respond to the constraints faced by smallholder farmers—and particularly by women—not only in terms of accessing land, seeds, credit and other productive assets but also in terms of marketing produce and getting a fair price.

In identifying these longer-term interventions, understanding of the problem fails to acknowledge the systemic economic, social and cultural causes of food and nutrition insecurity that are inherently gender-unjust. They fail to take into account the enormous yet often invisible contribution women and girls already make to food and nutrition security in the form of unpaid productive and care work. As a result, solutions place too much emphasis on enhancing the flow of productive resources for women. They do not well-address gender-injustice in transformative ways that challenge and transform the gender-inequitable distribution of food at the household level, tackle the issues of GBV and women's unpaid care work, and provide opportunities and choices for women and girls that go beyond the domestic realm. This compromises progress toward both gender equality and food and nutrition security, which are inextricably linked.

To develop policies and programs that promote gender justice as both a means to food and nutrition security and an end goal, it is vital to have a clear vision of what success should look like. IDS (2015) *op. cit.* suggests what this vision could be, and that progress toward its achievement needs to be mapped through progressive, gender-sensitive indicators. The IDS paper lists the core principles that must underpin thinking and action on gender-just food and nutrition security, and recommendations for translating these principles into practice. A good example is Oxfam's Women's Economic Leadership initiative. The production of participating women farmers has increased, but they have also acquired more decision-making power and their confidence and skills have grown through working in women's agricultural collectives. Sisto (2007) points out that to adequately design gender-responsive food security programs, one needs first to know the demographic, social, economic, political, institutional and physical security factors in a given area, and design actions accordingly. He gives a detailed breakdown of these in his paper.

Integrating gender equality concerns has assumed steadily growing importance in the work of the Asian Development Bank (ADB). In April 2013, ADB approved the *Gender Equality and Women's Empowerment Operational Plan, 2013–2020: Moving the Agenda Forward.* The plan emphasizes the need for deepening gender mainstreaming and for direct investments in women and girls to close remaining gender gaps and achieve better gender equality outcomes. Investments are needed in women's and girls' education, health and economic empowerment, as well as in public transport facilities, better water services and clean energy sources—especially in rural areas. The plan highlights the importance of supporting the efforts of ADB's developing member countries to establish and strengthen social protection strategies and programs that integrate gender dimensions in their design and implementation.

Measures that only help relieve women of their burdens and recognize their largely under-valued contributions to household chores and the "care" economy are insufficient. Such measures must be linked with strategies that pave the way to transformed equitable gender systems.

How best can social protection support access to food for low-income households—especially those headed by women—when incomes are insufficient for adequate living standards? The specific impacts of existing government social security programs on women and on gender equality remain largely ignored. Since these programs typically do not acknowledge the specific situation of women, women usually do not benefit from them as much as they should. Additionally, the opportunity these programs represent for gender empowerment may be missed. In its paper, ADB considers four tools through which social protection can be transformative for women. ADB is anxious to promote nutritional security too. Improved food intake translates into better health and nutritional outcomes only if accompanied by access to adequate education, health services, water and sanitation. Gender-sensitive policies that empower women within households can contribute to nutritional security. Provision of child-care services and the redistribution of power within the household are vitally important—not only to allow women to make the choices and provide the care that matter for infants but also to ensure that men value and contribute to such care.

Assessment of the agriculture–food–health continuum reveals that investments in agriculture aimed at improving productivity and access to markets, while essential components of food security strategies, will not necessarily translate into improved health or nutrition. Such investments should have the complementary aims of reducing micronutrient deficiencies through promotion of sufficiently diverse and balanced diets, and reducing rural poverty. The gender dimension is critical to addressing these links and to the effectiveness of support to agriculture. Support needs to increase the incomes of the poorest households, and within these households, benefit women in particular. Support also needs to incorporate an expanded role for women in decisions regarding the priorities of agricultural research and development.

#### 6.2.4.4 FEMINIZATION OF POVERTY

Womankind being held in check in the ways described earlier is in large part responsible for much of the food insecurity and undernutrition in the developing world, through the entrenched poverty that it brings about. Poverty was the first of the eight causes of food insecurity mentioned in Chapter 3 "Causes of Food Insecurity". Box 6.2 provides a quote from an FAO report discussing the socio-economic reality of rural women's status in much of the developing world.

One of the best-known and most impactful NGOs working to relieve poverty is Brac (formerly known by the expanded acronym—the Bangladesh Rural Advancement Committee). Some 150 million people have been assisted out of poverty by this organization. Once he had founded Brac in 1972, Sir Fazle Hasan Abed focused on the social and economic empowerment of women, which at the time, was a radical departure from the conventional approach. On his being awarded the World Food Prize in 2015, Sir Fazle commented: "The real heroes in our story are the poor themselves and, in particular, women struggling with poverty. In situations of extreme poverty, it is usually the women in the family who have to make do with scarce resources. When we saw this at Brac, we realized that women needed to be the agents of change. Only by putting the poorest, and women in particular, in charge of their own lives and destinies, will absolute poverty and deprivation be removed from the face of the Earth" (BBC News, 2015).

**Box 6.2 Rural Women's Invisibility**

*Despite their contribution to food security, women tend to be "invisible" actors in development. As a result, their contribution is poorly understood and often underestimated. There are many reasons for this. Work in the household is often considered to be part of a woman's duties as wife and mother, rather than an occupation to be accounted for in the national economy. Outside of the household, a great deal of rural women's labor—whether regular or seasonal—goes unpaid and is, therefore, rarely taken into account in official statistics. In most countries, women do not own the land they cultivate. When land is owned by women, it tends to be smaller, less valuable plots that are also overlooked in statistics. Furthermore, women are usually responsible for the food crops destined for immediate consumption by the household, that is, for subsistence crops rather than cash crops. Also, when data is collected for national statistics, gender is often ignored or the data is biased in the sense that it is collected only from males, who are assumed to be the heads of households.*

*Rural women's invisibility is further accentuated by their lack of political power and social representation resulting from prevailing attitudes, gender-biased legal and social structures and illiteracy, among other factors. Extension services reach women much less frequently than they do men.* Statistics indicate that women receive no more than 5 percent of extension resources. *This lack of knowledge often hinders the progress of women and their contribution to food security, particularly at the family level.*

*The combined effect of these handicaps is an increasing feminization of poverty.* Since the 1970s, the number of women living below the poverty line has increased by 50 percent, in comparison with 30 percent for their male counterparts. More than 70 percent of the 1 300 million poor people today are women.

**Source:** ***Retrieved from 'Women Feed the World', www.fao.org/docrep/x0262e/x0262e16.htm (accessed June 14, 2015). Reproduced with permission from FAO.***

## 6.3 LAND TENURE

### 6.3.1 Introduction

Land tenure is a complex social institution which governs the relationship among people with regard to assets such as land, water bodies and forests. It can have a legal or customary basis, or both. Access to land for the rural poor is often based on custom rather than title deed. Land tenure regulates use, control and transfer rights, whereby an individual or group, or government, owns a parcel of land and whether that land can pass through inheritance to children of those who currently use it. It determines the rights of individuals to make money from the land through agricultural means, building on it, leasing or selling it to another party, and who has the right to declare that others do *not* have rights to use that parcel of land. Many a war has been fought over competing claims to land tenureship, and disputes are numerous at the international level—witness the islands in the South China Sea, for example.

Very often there is lack of clarity surrounding land ownership. For instance, the opportunity in Zambia to plant more of the fertile land is fraught with land reform difficulties, with respect to the disconnect between the customary rights of some 70 tribes and the "legal status" of the land, which belongs to Government. And in West Bank Palestine, in Burin village in

Nablus governorate for instance, a piece of land may be registered in the name of someone who died long ago or has become a refugee, and for a descendant using that land today to prove ownership by right, he has to demonstrate to the Village Council that none other of perhaps 50 theoretical claimants have, or intend to exercise, a challenge to the occupant's right of use.

As another example, land ownership and use in Liberia is extremely complicated. Multiple systems of land use and ownership exist side-by-side, complicating the resolution of land disputes. Ownership systems have been imposed by various regimes over time, and any one piece of land can be conflicted by overlapping claims. Many Liberians still distrust the formal system, seeing courts as expensive timewasters, and women and young people are dissatisfied with traditional fora where gender and age likely count against them. With many Liberians returning to their lands after the civil war only to find them occupied by others, mediation offering engagement with both traditional and national institutions has been a welcome tool for peaceful dispute resolution among the various plaintiffs. The national Land Commission and the Norwegian Refugee Council have been instrumental in this task, in Nimba County on the border with Ivory Coast, for example. Ethnic tensions in Nimba that were fueled and exploited during the wars remain, and different groups have incompatible narratives of recent history. There are many disputes between Mano, Gio and Mandingo people over land ownership. Until these disputes are resolved, individual and national food security cannot be secured, against a backdrop of community trust having been destroyed by ethnic conflicts.

As a general point, unless a farmer or livestock keeper, or a community of such recognized in customary law at least, has ownership or security of tenure of a given piece of land, that person will not invest in maintaining its fertility, improving its grazing, or applying best water management practices on its slopes to maintain productivity and reduce erosion.

There is a strong gender component to land ownership, as discussed in Section 6.2. That a woman should be dispossessed of land she has been using because her husband dies, by his relatives taking over the land, seems to outsiders to be hugely unfair and discriminatory. Yet the cities of Delhi in India, and Sana'a in Yemen, for instance, have many women and their children living on the streets, having been thrown off their land because of such social mores. Landlessness can also come about as a "failing to cope with drought" mechanism, where the land used for the farming livelihood has to be sold to buy food for "today" (see Section 4.1).

Smallholders often seem to have any rights they have ignored in instances where governments have done so-called "landgrab" deals with offshore companies or States, in which they are granted the franchise for capital-intensive land development programs. Amazonia is a case in point (see Case Study 5 on the book's companion website), and oil palm plantations in Sarawak, Malaysia.

### 6.3.2 Land Tenure, Land Rights and Land Reform

A valuable discussion paper in 2010 links land tenure with the deeper issue of land rights (Wickeri and Kalhan, 2010). Even though land constitutes the main asset from which the rural poor are able to derive a livelihood, millions of such farming families do not enjoy ownership rights over it and are considered landless. Up to a quarter of the world's population is estimated to be landless (UN-Habitat, 2008).

An estimated 45% of the world's population still makes its living primarily from agriculture. In many developing countries, and a number of transition economies, these agricultural families still constitute a substantial majority of the population. However, many of these agricultural families lack a stable and predictable relationship with the land they farm, and thus face serious economic and social food insecurity. They earn their living mainly as tenant farmers or agricultural laborers. Tenant farmers typically pay high rents and have little security of land possession from season to season, whereas agricultural laborers (often itinerant) generally work for very low wages.

In developing countries, these landless 100 million farming families (equivalent to half a billion people) are among the poorest on earth. They constitute majorities or near-majorities of the agricultural population in countries like India, Bangladesh, Pakistan, Indonesia, the Philippines, South Africa, Brazil, Colombia, Guatemala and Honduras. In addition, they comprise a significant part of the agricultural population in many other countries, from Zimbabwe and Egypt to Afghanistan, Nepal and Venezuela. How to improve and secure the relationship which agricultural families have with land, is the central tenet of "land reform"—how to increase the ability of the rural poor and other socially-excluded groups to gain access to land and exercise effective control over it (Prosterman and Hanstad, 2003).

In some countries there have been redistributive land reform moves, to spread ownership of large blocks of land across a greater number of poorer people. Such has occurred in Zimbabwe, Pakistan and Brazil, and there is a strong demand for it from the poor in the Philippines (see Section 6.7.2), and Paraguay (where 80% of the land is owned by 2% of the people, and although legislation toward this goal has been enacted, it is poorly enforced). The methods used to achieve land reform can be controversial, varying from mild to robust, and there can be unfortunate outcomes.

*According to FAO, rural landlessness is often the best predictor of poverty and hunger.* While not the only path out of poverty, access to land helps rural households save money by growing their own food, while generating income through the sale of crops and livestock (de Janvry and Sadoulet, 2001). Secure land ownership avails access to credit markets, encourages investment to improve its productivity and enables wealth to be passed to the next generation. Land also provides a valuable safety net as a source of shelter, food and income in times of hardship, and a family's land can be the last available resort in times of disaster. Moreover, access to land avails a range of other fundamental human rights, including those to water, health and work.

The International Bill of Human Rights comprises the Universal Declaration of Human Rights (adopted by the UN General Assembly in 1948), the International Covenant on Economic, Social and Cultural Rights (ICESCR), and the International Covenant on Civil and Political Rights, the latter two both coming into force in 1976.

Article 11 of the ICESCR says as point 2: "The States Parties to the present Covenant, recognizing the fundamental right of everyone to be free from hunger, shall take, individually and through international cooperation, the measures, including specific programs, which are needed: (1) To improve methods of production, conservation and distribution of food by making full use of technical and scientific knowledge, by disseminating knowledge of the principles of nutrition and by developing or reforming agrarian systems in such a way as to achieve the most efficient development and utilization of natural resources; (2) Taking into account the problems of both food-importing and food-exporting countries, to ensure an equitable distribution of world food supplies in relation to need".

The nine-page General Comment 12[7] (The right to adequate food) of 1999 about Article 11 of the ICESCR gives much more detail about the right to land and food than that given in Article 11 itself.

### 6.3.3 Land Tenure of the Pastoral Afghan Kuchi

Case Study 7 on the book's companion website, provides a snapshot of one of the world's pastoral livelihoods peoples, the Kuchi of Afghanistan. In addition to problems related to *grazing rights and access to pasture*, which are most prevalent in the summer areas, issues related to *residential* land insecurity exist. The winter area is, in relative terms, the more permanent base for the Kuchi. Increasingly, communities are settling on land where they would traditionally dwell only in the winter (and to a lesser extent in the spring or summer). Destitute Kuchi having lost their livestock to drought or some other calamity are the first to settle. Yet, even among the migratory, livestock-dependent pastoralists, there is a tendency for communities to move toward a system of partial migration. These partially migratory communities comprise households that migrate in the spring, and others that stay behind.

Local residents will allow the presence of the Kuchi on the land, as long as this is on a seasonal basis. All is well so long as they pitch their tents for a few months a year. But establishing a well (never mind a school, community health center or veterinary clinic) is quickly associated with the Kuchi

[7] Retrieved from http://www.refworld.org/docid/4538838c11.html (accessed June 10, 2015).

increasing their customary claim to the land, which evokes a negative reaction in the local residential population. Such land tenure insecurity hampers modernization of the "Kuchi way of life".

Plowing up of pasture land for rainfed agriculture is an increasingly widespread phenomenon in Afghanistan. In the hope of getting a quick return, the indigenous vegetation is destroyed, often leading to severe erosion. For instance, FAO has estimated that 50% of Dasht-e-Laily in the north has been plowed up, a process that started probably in the early 1990s and exacerbated during recent years. Low livestock numbers and the fleeing of many pastoralists from the area led to a vacuum, which led to occupation of those areas by commanders. A similar story applies for the Shiwa area in Badakhshan, where an estimated 22% of pasture has been converted into agricultural land (Patterson, 2004). In addition to a reduced grazing area and environmental degradation, this sequence of events leads to a shift in control over the pastures from local, livestock-dependent people to the powerful elite. The result is disempowerment of the poorest, as it is often the poorest (settled or Kuchi) who are most reliant on pasture for their livelihood.

One of the main constraints on the establishment of a non-pastoral livelihood is residential land tenure insecurity in areas where they settle, mainly but not exclusively in the winter areas. Therefore, while residential land tenure insecurity is a constraint to modernizing the "pastoral way of life", for those destitute Kuchi who have dropped out of pastoralism, the problem is even more daunting.

Settled or partially settled Kuchi communities often continue to live in tents or very simple structures, because the settled people will not allow more permanent building. Many cases are known where Kuchi are subjected to extortionate random taxing by local commanders or villagers, and are constantly at risk of being evicted. This leads not just to inability to rebuild a livelihood, but to extreme insecurity, especially for women. They tend to be the first to plead for a more sedentary lifestyle, as they have the task of having to walk long distances to the nearest water source, and packing up the tents and pitching them elsewhere.

The current state of insecurity, over both residential land tenure security and access to resources that the livelihood depends upon, disempowers the Kuchi. Whatever demand-led interventions the Afghan government and partner organizations may seek to implement in both pastoral and nonpastoral livelihoods, success can come only if built upon a solid foundation of tenure security. The establishment of this is a pre-condition for any further investment in these livelihoods. In the process, a distinction needs making, through (possibly facilitated) negotiation and agreement, between *common property* (eg, pasture reserved for the resident community, to which possibly access rights can be awarded to other communities) and *public land* (eg, pasture that can be used by all, according to agreed-upon access rights for specific communities).

## 6.4 POPULATION GROWTH PRESSURE ON LAND AND WATER RESOURCES

There is today a challenge like never before to the sustainability of land and water resource management. It is a truism that the basis of feeding the world depends on agricultural productivity of the land, which itself primarily depends on water availability and soil fertility. As discussed elsewhere in this book, the world is only just managing to feed the seven billion people on the planet now, and 795 million of those are not well-fed but hungry. FAO (2011a) has identified parts of the developing world where land and water resources will be least able to cope with increasing demand put on them in the years ahead. The global community needs to focus on these "systems at risk" and adopt concerted and timely remedial interventions, through investments and international cooperation, not only on a global scale but locally, where the consequences of lack of action on agricultural livelihoods are likely to be greatest.

Annual rates of growth in agricultural production have been slowing, now only half of the 3% witnessed in developing countries in the recent past. The growing competition for land and water is now thrown into stark relief as sovereign and commercial investors begin to acquire tracts of farmland in developing countries (see Sections 3.8 and 6.8). Production of feedstock for biofuels competes with food production on significant areas of prime cultivated land (see Section 3.9), and climate change brings another raft of challenges (see Section 6.9).

Structural problems have also become apparent in the natural resource base. Water scarcity is growing, as irrigation systems expand and demand from industry and domestic use rises. For instance, the Huang He (Yellow) River in China no longer reaches the sea year-round, and the Chinese government estimates that around two-thirds of its water is too polluted to drink. Lake Chad in West Africa is a shadow of its former self, old irrigation channels orphaned above even the annual high water mark of the lake. Globally, groundwater is being over-pumped, and aquifers are becoming increasingly polluted and salinized in some coastal areas, such as Libya. Agricultural production per unit of water will become as important as production per unit of land, as will the development of integrated water management plans.

Overall, the picture is also that of a world with *increasing imbalance between availability and demand for land and water resources* at local level: the number of regions reaching the limits of their production capacity is fast increasing. This applies not just to the "usual suspects" like the rainfed semi-arid areas of Africa. Irrigated rice-based systems in Asia are also at risk, with little opportunity for intensification or expansion in the face of population growth and ecosystem degradation. Another of the "at-risk" regions identified in the FAO (2011a) Report *ibid.* comprises deltas, coastal alluvial plains and small islands. These are under threat from industrial pollution of water courses, rising sea and increasing frequency of severe weather.

In an earlier book that this author wrote on food crops and drought, the pre-publication reviewer pointed out an omission in the text—that there had been scant mention of reducing the rate of human population growth—"Unless this can be brought about, everything we scientists do to produce more food will come to naught". Justly chastised, the omission was corrected, and it is pertinent now to repeat the cogent quantitative caution from 1997 by Dr Alistair Allan that was included in that book, with reference to Kenya, then suffering from a deep drought (see Box 6.3).

**Box 6.3 Demographic Pressure on Resources in Kenya**

"In 1967, there were only 9 million Kenyans; now there are 31 million. Even if there were never again a failure of the rains, food production could not increase sufficiently to cater for the current annual increase in population of about a million, and the disparity between food production and consumption is set to widen further".

"Allan also points to the daunting practical implications of such population increase in other national sectors. For instance, 1 million extra children a year need that number of extra school places made available. At, say, 1,000 pupils per school, an extra 1,000 schools need to be built each year, or three per day, with accompanying teachers and equipment. This is an impossible target for Kenya to achieve".

[Fortunately, Kenya is making progress on reducing the rate at which her population is increasing, now at an annual rate[8] of 2.7%.[9] Other countries in Africa are doing well too, some even better than Kenya—see below].

Successful address of current population growth in the developing world is a major means of reducing the unsustainable pressure on the planet's finite land and water resources, which threatens the well-being of those who inhabit it. On June 13, 2013, in New York, the United Nations Population Division released its comprehensive estimates and projections, *World Population Prospects: The 2012 Revision.* The then current world population of 7.2 billion was projected to increase by 1 billion by 2025 and reach 9.6 billion by 2050, up from the 9.3 billion that the UN projected in its 2010 *Revision.* A major reason for the higher projection is higher fertility (birth rates) in some countries than previously estimated, particularly in Africa. For example, in 15 high-fertility countries of sub-Saharan Africa, the estimated average number of children per woman has been adjusted upwards by more than 5%.

Population increase will be mainly in developing countries, with more than half in Africa. The report notes that the population of developed regions will remain largely unchanged at around 1.3 billion from 2013 until 2050.

[8] This is globally defined as the average annual percent change in population, resulting from a surplus (or deficit) of births over deaths and the balance of migrants entering and leaving a country.

[9] Retrieved from http://data.worldbank.org/indicator/SP.POP.GROW (accessed April 18, 2014).

By contrast, the 49 least developed countries are projected to double in size from around 900 million people in 2013 to 1.8 billion in 2050. While there has been a rapid fall in the average number of children per woman in large developing countries such as China, India, Indonesia, Iran, Brazil and South Africa, rapid growth is expected to continue over the next few decades in countries with high levels of fertility such as Nigeria, Niger, the Democratic Republic of the Congo, Ethiopia and Uganda, and also Afghanistan and Timor-Leste, where there are more than five children per woman. Actual fertility rates over the next few decades could of course change these predictions.

The report notes that India is expected to become the world's largest population, passing China around 2028, when both countries will have populations of 1.45 billion. After that, India's population will continue to grow while China's is expected to start decreasing. Meanwhile, Nigeria's population is expected to surpass that of the United States before 2050. Europe's population is projected to decline by 14%, the report states, with the continent already facing challenges in providing care and support for a rapidly aging population.

Rwanda is one of Africa's smallest nations, and also one of its most densely populated, with over 455 people per square kilometer on average, according to the Government's website, based on demographic data prior to release of results from the census of August 2012. It is clear that unrestricted population growth is a "no no" for the country, which has bounced back from the 1994 genocide in exemplary fashion and wishes to sustain the well-deserved status of an African success story. In 2005, increased demand for family planning in Rwanda was evident in the results of a Demographic and Health Survey. Although women were having an average of 6 children each, the reported desired family size was 4.3. Nearly 4% of married women did not wish to become pregnant, yet were not using contraception. Hence a strong "demand" had been identified. Worldwide, increased contraceptive use is a result of one or two basic interventions: demand and/or supply programs. In other words, women and/or men adopt contraception either because they feel they need contraceptives (demand side) and/or because contraceptives are easily accessible (supply side).

Recognizing that population growth is the major barrier to achieving the ambitious 2020 Rwandan Vision for Development,[10] President Kagame and his Minister of Health have taken the lead in supporting and encouraging family planning (Madsen, 2011). To increase family planning coverage, the Rwandan government increased the budget for family planning activities and extended the number of partners, initially represented only by USAID and UNFPA. Between 2004 and 2007, the budget was increased sixfold, rising

[10] The general objective of the 2020 vision is to transform the country into a middle-income country by the year 2020 (Rwanda Vision 2020. Kigali, July 2000. Ministry of Finance and Economic Planning. 31 pp.).

from 91,231 USD to 5,742,112 USD (Ministry of Health, 2009). The main partner was the *Twubakane* project, which played a major role by supporting the government in various trainings for providers and construction of secondary health posts *inter alia*.

Other strategies in Rwanda include strengthening health facilities with a national network of *mutuelles*—innovative community-based health insurance schemes supported by member premiums and government funding, and conducting mass media campaigns (with the message that besides making poverty reduction easier, having smaller families also increases health and education opportunities). These strategies were instrumental in making contraceptives more widely available and affordable. The joined-up multi-strategy approach, enormous number of players involved and the high level of commitment has been documented (Solo, 2008). The success of family planning initiatives in Rwanda, using statistical analysis, albeit on an acknowledged incomplete range of factors, has been documented to identify those most responsible for the achieved success (Muhoza et al., 2013).

Rwanda has exhibited a dramatic increase in contraceptive use over recent years. The contraceptive prevalence rate (CPR) has increased three-fold, from 17% in 2005 to 52% in 2010. This achievement has surpassed Rwanda's national objectives, as recorded in the 2008–12 Economic and Poverty Reduction Strategy. This increase in CPR was accompanied by a large decline in unmet demand for family planning, from 38% to 19%, and a decline in the total fertility rate (TFR), from 6.1 to 4.6 births per woman between 2005 and 2010 (National Institute of Statistics and ORC Macro, 2006; National Institute of Statistics et al., 2012).

No other East African Community country has recorded a comparable achievement (see Table 6.1). Elsewhere, a rapid fertility decline of similar scale to that in Rwanda has occurred in only a few other developing countries, including Cuba, Iran, Mauritius, Algeria and parts of East Asia, but Rwanda is without peer in this regard in sub-Saharan Africa (Madsen, 2011) *ibid*.

Despite the recent rapid decline in fertility rates and increased use of family planning, Rwanda remains a high-fertility country, and demographic

**TABLE 6.1** Changes in Contraceptive Prevalence Rate (CPR), Unmet Need, and Total Fertility Rate (TFR) in Rwanda, Uganda, Tanzania and Kenya Over Recent Times

| Indicator | Rwanda 2005→10 | Uganda 2006→11 | Tanzania 2004/05→09 | Kenya 2003→2008/09 |
|---|---|---|---|---|
| CPR (%) | 17.1→51.6 | 23.7→30.0 | 26.4→34.4 | 39.3→45.5 |
| Unmet Need (%) | 37.4→19.2 | 40.6→34.3 | 21.8→25.3 | 24.5→25.6 |
| TFR | 6.1→4.6 | 6.7→6.2 | 5.7→5.4 | 4.9→4.6 |

*Source: Reproduced with permission from Muhoza et al. (2013).*

momentum will drive continued population growth. Even if average family size falls below four children per woman, the population is projected to nearly double over the next 25 years, so there should be no let-up in the family planning initiatives if food security and living standards are to be protected in a sustainable manner.

A review of findings from many studies found that family planning programs have been most successful when they have used a variety of approaches, mixing those that improve the quality of services with those that address sociocultural barriers or that focus on winning community and social support for family planning use (Mwaikambo et al., 2011).

An analysis by the Acquire Project (2005) on the success of family planning programs in Malawi, Ghana, and Zambia showed that family planning programs in these three countries were successful not just through supply-side interventions, but also through effective and innovative efforts on the demand side, including both working with the community and bringing services closer to rural populations. The messages that were effective in creating demand and changing behavior were developed in consultation with the community to ensure that they were appropriate and meaningful.

An excellent case of contraception supply-side initiatives is in Bangladesh, where the family planning program has promoted the distribution and sensitization for family planning by women themselves, overcoming the cultural norms that subordinate women. Contraceptive use has increased substantially among women with little education, as well as among women of higher socioeconomic status, leading the three authors of a report on the success to suggest that the family planning program reorient itself to target men (Schuler et al., 1995).

According to World Bank in 2011,[11] in Latin America and the Caribbean, fertility rates have been declining since the 1960s, when women had an average of six children, although, paradoxically, the population has tripled. Among other factors, this increase is due to improved healthcare services and rising life expectancy in the region. "The population continues to grow, but at a much slower pace than it did a century ago. By 2050, the growth rate is expected to approach zero and the population will stabilize at 800 million, 8% of the projected global population" said Joana Godinho, the World Bank's human development sector manager for Latin America. For Godinho, the region is still experiencing a demographic benefit since most of the population is economically active. However, the population will rapidly age given the declining fertility and death rates, just as has occurred in developed countries. "This has an impact on public spending in health and pensions, as well as on poverty, inequality and economic growth" said Godhino, who stressed that the level of impact will depend on government actions to address the change. Some measures to prepare for the region's new demographics include the promotion of healthy lifestyles and lifelong learning for a long productive life.

[11] Retrieved from http://go.worldbank.org/I6VGAJX960 (accessed April 6, 2015).

## 6.5 EMPOWERMENT AND RESILIENCE THROUGH ACCESS TO FACTORS OF PRODUCTION

The African Union through its Comprehensive Africa Agriculture Development Program (CAADP)[12] has identified *four challenges to food security* (1) inadequate risk management at all levels from household to regional levels, (2) inadequate food supply and marketing systems for distributing food, (3) lack of income opportunities for the vulnerable, and (4) hunger and malnutrition. To overcome these challenges, a *Framework for African Food Security (FAFS)* has been developed with the following objectives: (1) improved risk management, (2) increased supply of affordable commodities through increased production and improved market linkages; (3) increased economic opportunities for the vulnerable, and (4) increased quality of diets through diversification of food among the target groups.

The narrative stresses the need for efficient farming, selectively using sustainable modern technologies to improve profitability through improved productivity (per unit of land, water, labor and unit of investment), on either smallholder or larger scale. Even the latter can be sustainable, associated with individuals or groups who/which have a concentration of capital, which they invest in appropriate technologies, developed as a result of research, and can afford to take risks with climatic and other uncertainties which low-risk smallholders cannot do. In the process, huge incremental employment can result, with money availed to people who used to be subsistence smallholders, enabling them to be more food-secure, improve their living standards, re-roof their houses when needed, better-educate their children, invest in their own farmland to raise its productivity and profitability, improve the household's food self-sufficiency, and create paid work on the land for the family or the local community.

Moreover, at the national level, money accruing to government through taxes of various sorts from the enterprise owners and workers in it, can be used to provide better public services and enable nations to withdraw gradually from the requirement for international assistance (need for which is highly competitive and may lead to donor-driven agendae). The global fund for international largesse is reducing because of domestic austerity measures and the ever-increasing demands made on it because of natural and manmade crises.

Marketing can be improved, with better-organized local involvement through cooperatives and service centers, and the reduced role of exploitative middlemen such as afflicts the smallholder cocoa farming sector in northern Liberia, for instance. The result can be a fairer price achieved for local farmers and hence more money to invest in local communities, to improve productivity and resilience to adversity. In this way, smallholder communities are empowered, and enabled to improve without losing control of their livelihoods. The private sector (through providing physical inputs, credit, finance,

[12] Retrieved from www.nepad-caadp.net (accessed March 5, 2015).

markets, etc.) is then helping them rather than exploiting them. A successful model of this has existed for decades in the Kenyan rift valley, centered on Nakuru, with smallholders becoming wealthy because of the free market in inputs and agricultural produce. The process is under way in Zambia also, with conservation farming technology being adopted which is transforming smallholder yields and effecting better gender equity (see Section 6.2, and Case Study 2 on the book's companion website).

## 6.6 FOOD SECURITY GOVERNANCE

*Good governance is perhaps the single most important factor in eradicating poverty and promoting development*

**Kofi Annan, former UN Secretary-General (1997–2006), at the 4th High Level Forum on Aid Effectiveness, Busan, Korea, 2011**

### 6.6.1 Meaning of Governance

Food insecurity is caused by a complex interplay of factors—technical, social, economic, institutional and cultural. While some are outside of the direct control of governments, there is recognition that institutions, laws, regulations and political processes play an important role in either enabling or constraining particular pathways to all four pillars of food and nutrition security, and better livelihoods and well-being for all (see also Section 3.6).

In its original interpretation, the term "governance" had a narrow meaning—the manner of conducting Affairs of State, being almost synonymous with "government". In more recent times, "governance" has been used to describe the process of decision-making and implementation or otherwise of decisions and policies, the way power is distributed and exercised within a society—especially national and local, international, institutional, public and private sector and civil society, and interactions among them. Democratic governance is a common term now referring to the *political* dimension of governance, and its degree of legitimacy.

Various international organizations such as the IMF, UN Commission on Human Rights, World Bank, UNDP and the European Commission have variant definitions of "governance" (FAO, 2011b). There are common principles though across the range. Six overarching governance components suggest themselves as most likely to improve the chance of sustainable and equitable gender-sensitive advances in the country concerned: efficiency and effectiveness of the public sector, equality and fairness, accountability, participation and responsiveness, transparency, and finally, the rule of law.

In the food and agriculture area, the concept of good governance is increasingly applied at the *sector* level, for instance land and water governance, fisheries governance, forest governance and so on. Good governance of *food and nutrition security* has the potential to ensure that food security programs are effectively implemented, to the greatest benefit of those who most need them. When governance structures, both formal and informal, abide by the rule of law, exercise their functions in a responsive and equitable manner, and give voice to a wide range of diverse interests, including those of the food insecure and hungry, resulting activities should contribute to improving food security in a country and globally.

A key lesson learned from examining the experiences of countries attempting to meet the MDG 1c target is that hunger, food insecurity and undernutrition are complex problems that cannot be resolved by a single stakeholder or sector. Addressing the immediate and underlying causes of hunger requires a variety of actions across a range of sectors, including agricultural production and productivity, rural development, forestry, fisheries, social protection, and trade and markets. While many of these actions will be at national and local levels, there are also issues of a regional and global nature that require action on a wider scale. A major task incumbent on food security governance is to foster an "enabling environment" that will create incentives for all sectors to improve their impact on hunger, undernutrition and food insecurity. Commitment is needed at the highest level, including provision of adequate financial and human resources.

### 6.6.2 An Integrated and Coordinated International Approach

At the *international* level, the apex institution which is food and nutrition security outcome-oriented is the FAO Committee on World Food Security (CFS).[13] Indeed, the need for global governance of food led to the creation of FAO in 1945. CFS is the coordinating body on food security, comprising an inclusive international and inter-governmental platform to promote the elimination of hunger and ensure food security for all. CFS is the United Nations' forum for reviewing and following up on policies concerning world food security. It also examines issues which affect the world food situation. CFS was established as a result of the food crisis of the 1970s, on the recommendation of the 1974 World Food Conference.

Whether or when sustainable food security and nutrition security are achieved in the developing world to a large extent depends on what happens in the developed world, particularly how free markets work and are regulated. There is also a "tension" between crop and livestock production, which requires monitoring. With nearly two-thirds of the EU's cereal production used for animal feed, only around one-third is for direct human consumption. Dietary change in developing countries toward more meat and dairy products will influence food availability and access for future generations.

[13] Retrieved from http://www.fao.org/cfs/cfs-home/en/ (accessed March 1, 2015).

It will be more difficult for the poorest among the expected global population of 8 billion in 2030, and more than 9 billion in 2050, to eat sufficiently if diets become richer than now in livestock and dairy consumption.

This in part is because livestock will increasingly compete with humans to consume the world's grain production, and also because population growth will increase competition for available land, water and inorganic fertilizer. This in turn will put an upward pressure on food production costs and food prices, making it more difficult for those on low incomes to afford an adequate diet, whereas people on high incomes will be able to choose their diet without concerns of affordability. There is also the fact that livestock which consume grain rather than pasture or browse represent a huge inefficiency compared with humans directly consuming that grain. Sheep and cattle have a low feed conversion ratio, needing 8 kg of feed to add 1 kg body weight. Poultry and fish are more efficient converters. Notionally, if less grain is diverted toward fattening livestock, the sooner the MDG 1c outcome figure of 795 million hungry people can be reduced.

Although FAO has creditably taken a major lead in promoting the eradication of hunger, it is not agriculture and incremental food production alone which can resolve the developing world's food insecurity challenge. Health, education, clean water and sanitation, commerce and industry, for example, all have equally valuable roles to play to ensure that the four components of food security are achieved. This needs to include well-informed and joined-up policies, strategies and legal frameworks from governments, implemented and monitored; public and private investments to raise agricultural productivity; better access to inputs, land, services, technologies and markets; measures to promote rural development; social protection measures to ensure nutritious food for the most vulnerable, including strengthening the resilience of their communities and livelihoods to shocks from conflicts, climate change and natural disasters; and, specific nutrition programs, particularly to address protein and micronutrient deficiencies in mothers and children under 5 years of age.

Regions such as Africa, Latin America and the Caribbean, as well as a number of individual countries have strengthened their political commitment to food security and nutrition. For instance, in 2013 at the first summit of the Community of Latin America and the Caribbean States (CELAC), Heads of State and Governments, representing nearly 90 States and over 1.5 billion people, endorsed the UN's 2025 Zero Hunger target by reaffirming a regional commitment to the *Hunger-Free Latin America and the Caribbean Initiative* to end hunger by 2025. Moreover, at the African Union summit in July 2014, in Malabo, Equatorial Guinea, African Heads of State committed to end hunger on the continent by 2025. Such commitment strengthens ongoing efforts within the framework of the Africa-led CAADP of the New Partnership for Africa's Development (NEPAD), as well as the decision to increase South-South Cooperation efforts within Africa, as signaled by the Africa Solidarity Trust Fund for Food Security, established in 2013.

### 6.6.3 Sustained National Political Commitment

At the *national* level, deficiency in the rule of law encourages high rates of corruption, and social and gender inequalities. Conflicts are often linked to unequal access to land and other natural resources. Inequalities of access to natural resources and to inputs and services such as seeds, fertilizers or credit strongly limit agricultural productivity. Poor governance diminishes the performance of a given sector institutions and actors, as well as the outcomes of policies.

Even a well-balanced portfolio of natural resource, social and economic policies will remain ineffective in the absence of effective systems for service delivery, regulation, control of corruption and underpinning rights. Lack of transparency and information about social assistance programs to possible beneficiaries, and wide administrative discretion opening the way to ethnic or political favoritism or discrimination, can lead to failure of safety net programs to reach those in greatest need. Lack of citizens' capacity to hold governments to account is common in many countries having a low level of food security. If government institutions in charge of policy formulation and implementation do not feel morally or legally compelled to heed the concerns of the general public, in particular the politically weak rural population, they will more likely respond to well-organized lobbyists if they take to the streets, as happened in the food riots of 2008.

While quality dimensions of good governance do not in themselves ensure that the pillars of food security are governed well, their absence severely limits that potential and can, at worst, impede it. Government commitment to tackle the complex issue of food security will unlikely develop where there is no organized, politically active and mobilized constituency pushing the issue higher on the public and political agenda. It helps if there is a national Food Security Strategy in place, formulated cross-sectorally and in a participative way, and endorsed by Cabinet (see Section 5.2).

## 6.7 GOVERNMENT CAPACITY TO FORMULATE, IMPLEMENT, AND INSTITUTIONALIZE CHANGE AND REFORM

### 6.7.1 Identifying the Need for Policy Change

Introspection prior to action on making national planning, policies and strategies more evidence-based is a necessary first step toward food security framework reform. In some countries, such as Columbia in South America, the need to consider policy adjustment has been lent urgency by farmers' concerted protests over recent years. Since 2013 in Columbia, a surge in grassroots radicalism has effectively shut down the country through civic strikes. Farmers, mobilized under the umbrella organization *Dignidad Agropecuaria Colombiana* (Colombian Agricultural Dignity), say that government's failure to implement most of the policy reforms agreed with farmers in 2013 had turned small-scale farming into a loss-making business, and driven them to bankruptcy. The farmers blame the free trade agreements with the United States and the EU, of 2012 and 2013 respectively, which have led to the market being flooded with agricultural produce at prices which Columbian farmers cannot match. There are strong calls for food sovereignty (see Section 6.8). Indeed, the rise of neo-populist regimes throughout Latin America over the past two decades represents a challenge to the ability of the US government and its allies in the IMF, World Bank and WTO to advocate neo-liberal economic reforms as the solution to the region's chronic underdevelopment.

Listed below are five of the nine qualitative take-away messages from the 2013 edition of *The State of Food Insecurity in the World* (FAO, IFAD and WFP, 2013), these referring to the importance of *policies* (specifically mentioned in four of the points, and strongly implied in the fifth) (see Box 6.4 - italics not original):

**Box 6.4 Some Messages from The State of Food Insecurity in the World, 2013**

- Growth can raise incomes and reduce hunger, but higher economic growth may not reach everyone. It may not lead to more and better jobs for all, unless *policies specifically target the poor*, especially those in rural areas. In poor countries, hunger and poverty reduction will only be achieved with growth that is not only sustained, but also broadly shared.
- Undernourishment and undernutrition coexist in most countries. However, in some countries, undernutrition rates, as indicated by the proportion of stunted children, are considerably higher than the prevalence of undernourishment, as indicated by inadequacy of dietary energy supply. *In these countries, nutrition-enhancing interventions are crucial to improve the nutritional aspects of food security.* Improvements require a range of food security and nutrition-enhancing interventions in agriculture, health, hygiene, water supply and education, particularly targeting women.
- *Policies aimed at enhancing agricultural productivity and increasing food availability*, especially when smallholders are targeted, can achieve hunger reduction even where poverty is widespread. When they are *combined with social protection and other measures that increase the incomes of poor families to buy food*, they can have an even more positive effective and spur rural development, by creating vibrant markets and employment opportunities, making possible equitable economic growth.
- Remittances, which have globally become three times larger than official development assistance, have had significant impacts on poverty and food security. Remittances can help to reduce poverty, leading to reduced hunger, better diets and, *given appropriate policies*, increased on-farm investment.
- *Long-term commitment to mainstreaming food security and nutrition in public policies and programs* is key to hunger reduction. Keeping food security and agriculture high on the development agenda, through comprehensive reforms, improvements in the investment climate and supported by sustained social protection, is crucial for achieving major reductions in poverty and undernourishment.

*Source:* Reproduced with permission from *FAO, IFAD and WFP (2013).*

### 6.7.2 Participatory Driver of Change

Chapter 5 "Prevention of Future Food Insecurity" charted a number of actions that in their address could prevent the scourge of food and nutrition security. That these are not happening to the extent that they could in the affected countries often relates to the shortage of necessary knowledge and skills in government and national institutions. When such applies, capacity building at all levels of planning and implementation, from central and state governments to village volunteers is to be encouraged, thereby fostering improved human and social capital. Technical issues certainly feature strongly, yet as important is organization and management capability, vision and the strategic planning and policy making necessary to achieve that vision.

Across the developing world, *awareness campaigns* run by governments together with various international and national agencies and NGOs can have a strong impact on people's food and nutrition security, for the better. Such campaigns underlay the massive reduction of slum areas in Morocco for instance (see Case Study 10, para 10.2 on the book's companion website), and the resulting improved food security and nutrition indicators. Awareness campaigns on nutrition, health and hygiene, clean water and sanitation, to promote food and nutrition security are other examples whereby a demand can be created to speed the supply of available improved "technologies". Demands created which are not then satiated, is a recipe for social unrest of course.

There are many examples of a groundswell of rational initiatives from across society to improve the capacity of a country to address food insecurity, involving actors from public and private sectors, and civil society, working in harmony. The food price riots of 2008 are propelling governments of many countries to reform policies and change practices in support of national preparedness and building resilience (see Sections 5.1.3 and 6.5). 2015 could provide a window in which movement for change can happen rationally. The previous year had seen some bumper grain harvests, from the United States to Ukraine, and in much of Africa for instance, and oil prices and interest rates were low. Decisions made in 2015 may be more rational than those made by governments faced with rioting in the future.

Examples are given below of pressure which is already being exerted or is needed, on governments by their citizens to change policies which have a direct bearing on their food security—in India and the Philippines.

*India:* In India, famines have resulted in more than 60 million deaths over the course of the 18th, 19th and early 20th centuries, yet the last severe one was that in Bengal of 1943 in which up to two million people died. These famines have been ascribed to drought and colonial policy failures. The major reason for the subsequent disappearance of severe famine is the responsible foresight and engagement of the government in enabling the Green Revolution to occur there, and by facilitating construction of strategic grain reserves (see Section 5.5.1), together with improved transport systems, fair-price stores and labor-intensive public works in time of crisis. Famine need not follow drought. The government of India has ensured that these reserves are located strategically, that management systems and technical expertise are in place for their running (though there is room for improvement in this), and crucially that funding for all this is ring-fenced in National and State budgets.

Indian government agencies at State and Federal level were expected to hold an unprecedented stock of 90 million tonnes of rice and wheat when the annual wheat procurement drive ended in June 2013, up from 78 million tonnes a year previously.[14] Over 99% of this government grain was still stored in 50-kg bags in state-owned warehouses or in the open air. Such has been the practice for decades despite storage losses of 10% or more. Improvements are under way currently that could revolutionize both the modalities and technology of government grain storage. Many initiatives among States and at the national Food Corporation of India (FCI) are propelling this drive for reform. These schemes include Public Private Partnerships (PPPs) for construction of dozens of steel silo storage facilities, long-term contracts with builders of traditional *godowns* (warehouses) under a scheme called Private Entrepreneur Guarantee (PEG), and the large-scale introduction by companies of storage services using hermetically sealed silo bags to hold cereals procured by the State.

*The Philippines:* As one of the most populous countries in the world, the Philippines is marked by inequality and vulnerability, despite its middle income status. Farmers, the majority of whom are women, comprise 4 out of 10 poor Filipinos. The causes of hunger in the country are multiple. The southern region of Mindanao, hosting the country's Muslim minority, suffers from poverty and landlessness. Skewed land ownership patterns remain unsolved and continue to plague agriculture. The Asian NGO Coalition for Agrarian Reform and Rural Development (ANGOC)[15] estimates that 2.9 million small farms (having use of less than 5 ha of land) occupy slightly more than one-half of the total farm area, while only 13,681 medium-sized and large farms (of more than 25 ha in size) account for 11.5% of the total farmland. In most cases, the farmer-owner relationship is still feudal, and land ownership is concentrated among a few who are not so much interested in agricultural sustainability and productivity, but in controlling the use of their land and consolidating their political power in rural areas.

This situation is further complicated by armed conflict, this having a direct impact on food production. In addition to land, another productive resource in short supply is capital. Agricultural productivity and food production is declining. There are ever higher retail food prices and decreasing spending power with low income (whether as farmers or farm workers), extreme weather events, pests and disease. In addition, the population is

---

[14] Retrieved from http://www.davidmckee.org/2013/06/15/grain-storage-trends-in-india/ (accessed March 10, 2015).

[15] Retrieved from www.fao.org/docrep/w6199t/w6199t11.htm (accessed March 10, 2015).

growing, exacerbating the persistent hunger and food insecurity problems in the country.[16]

Poverty incidence is particularly high among landless agricultural workers and farmers cultivating small plots of land and in areas where the concentration of land ownership remains with a few prominent clans. *The 15 provinces with the highest land redistribution backlog, from the baseline year of 1997 to the present, contain 30% of the country's total poor population.*

Farmers and their bleak situation need significant help. They need government to legislate and promote an effective agrarian reform program, and an enabling policy environment that puts smallholder agriculture/family farms at its center. The government is trying to implement social justice measures such as the Comprehensive Agrarian Reform Law (1988), Indigenous Peoples' Reform Act and Fisheries Reform Code. However, the absence of an *effective* land redistribution program discourages Filipinos from relying on agriculture for a livelihood and undermines their capacity to feed the nation. To make matters worse, since 1981, the Philippine government has been pursuing a comprehensive and radical program of trade liberalization, the impacts of which have been devastating to the agriculture sector, especially to food producers. As a result, the government could do well to seek advice on how it might improve its performance in an holistic and coherent way to improve the lot of its people, to forestall further social unrest at home.

## 6.8 FOOD SOVEREIGNTY

### 6.8.1 Background

The term "food sovereignty" was first coined in 1996, by members of *la vía Campesina*, the International Peasant's Movement, at that year's World Food Summit. In 2007, a multi-stakeholder conference was held in Mali on the subject, involving about 500 delegates from more than 80 countries, with an agreed communique issued at the end, an excerpt from which is given in Box 6.5. The following year, an inter-governmental panel under the sponsorship of the United Nations and World Bank, adopted a concise toned-down definition, also in Box 6.5.

Without knowing the "political baggage" inherent in the term as portrayed in the Mali Declaration, yet being conscious of the "food price crisis" of 2008 which ensued from the global financial crisis of 2007–08, a newcomer to the term "food sovereignty" may think that the concept stemmed from the latter, and epitomized the need for a community or nation to be self-sufficient in its key foods, so its food security would not be jeopardized by price rises elsewhere in the world affecting import prices. In this scenario, not just availability of food could be assured in that community but also economic access to it by members of that community.

[16] Retrieved from http://focusweb.org/content/right-food-and-food-security-philippines-what-numbers-say (accessed March 10, 2015).

**Box 6.5 Two Contrasting Definitions of Food Sovereignty**

1. *....Food sovereignty is the right of peoples to healthy and culturally appropriate food produced through ecologically sound and sustainable methods, and their right to define their own food and agriculture systems. It puts the aspirations and needs of those who produce, distribute and consume food at the heart of food systems and policies rather than the demands of markets and corporations. It defends the interests and inclusion of the next generation. It offers a strategy to resist and dismantle the current corporate trade and food regime, and directions for food, farming, pastoral and fisheries systems determined by local producers and users.*

*Food sovereignty prioritizes local and national economies and markets and empowers peasant and family farmer-driven agriculture, artisanal - fishing, pastoralist-led grazing, and food production, distribution and consumption based on environmental, social and economic sustainability. Food sovereignty promotes transparent trade that guarantees just incomes to all peoples as well as the rights of consumers to control their food and nutrition. It ensures that the rights to use and manage lands, territories, waters, seeds, livestock and biodiversity are in the hands of those of us who produce food. Food sovereignty implies new social relations free of oppression and inequality between men and women, peoples, racial groups, social and economic classes and generations.....*

***– from the Declaration of Nyéléni, Forum for Food Sovereignty, Mali, March 27, 2007***

2. *Food sovereignty is defined as the right of peoples and sovereign states to democratically determine their own agricultural and food policies*

**The panel International Assessment of Agricultural Science and Technology for Development (IAASTD), sponsored by the UN and World Bank, 2008.**

Yet food sovereignty has taken on a wider meaning for many individuals and groups set up to fight global poverty. The term encompasses incrementally what is not addressed under normal use of the term "food security", namely, the political economy of where and how the food is produced, and how equitably this is done. The term "food sovereignty" has moved beyond national borders to address and target world markets and cartels, price-fixing, hoarding and multi-national food distribution companies, and not least large-scale land acquisitions (landgrab) purchase and lease in Africa by rich countries, in the interests of *their* future food security, which has provoked protests in Mali, Madagascar and elsewhere. In Mali for instance, a Libyan investment fund has signed a 50-year lease for 250,000 acres of prime rice growing land, and a Kuwait investment fund has long had an interest in large land blocks in Sudan (see also Sections 3.8, 3.9.4, 6.3.1 and 6.8.1).

Some of the exponents of this interpretation of food sovereignty go as far as to laud the concept of the subsistence smallholder and decry the trend in some counties toward farming larger blocks of land. The Mali Declaration movement proposes an alternative food system to transcend what it sees as the deep-seated social, economic and ecological contradictions of the global food economy. Supporters of the concept argue that nationally based food systems based on local markets and "peasant agriculture" represent the best means of combatting hunger and poverty in the global south, and that food sovereignty is an anti-colonial program. "Food sovereignty" of the Mali Declaration exponents identifies the State and capital as complicit in assorted inequities and injustices within the corporate food regime, centered on the alienation of consumers from producers. The definition looks as though it could be expanded to include the outlawing of GM crops and campaigning against monoculture, in Ecuador, for instance.

In 1999 the Venezuelan government wrote the concept of food sovereignty into its constitution, and began to take drastic steps to reduce imports. The country's agricultural system had seriously declined over the 20th century largely due to the emphasis placed on the petroleum economy, and paying for food imports with the oil wealth generated. At one point, the country imported more than 85% of its food from foreign sources. Yet, Venezuela was and is held hostage to price fluctuations on international energy markets, creating huge vulnerability to events it cannot control (Kappeler, 2013).

In September 2008 Ecuador also incorporated the term in its constitution, declaring food sovereignty as a strategic goal and governmental obligation. This later developed into a food sovereignty legal framework with the approval of the Food Sovereignty Law (LORSA) in 2009. To further develop this legal framework, the *Conferencia Plurinacional e Intercultural de Soberanía Alimentaria* (COPISA) was created in 2010, as a participatory organization responsible for drafting nine supplementary laws that support the LORSA.

Since then, with varying degrees of success, other countries (Bolivia, Nicaragua, Peru, Nepal, Mali and Senegal) have followed suit, integrating the "Mali Declaration" term and meaning into their constitutions, laws or policies (Burnett and Murphy, 2014). This seems either an expression of the left-of-center political regimes in the country concerned and/or (as in the case of Mali and Senegal) the fear of land-grabs by outside countries, to produce food for themselves, or biofuel crops, both of which are seen as threatening livelihoods of smallholders, undermining the ability of the host nation to feed itself and leading to environmental pollution from the intensive cultivation methods used.

The influence of the early *la vía campesina* movement of the 1990s may well have been in part responsible for the broadening of the early 1970s concept of "food security", from one largely about "food availability" to a concept involving affordability and nutrition (Aerni, 2011). The root of *la vía campesina* movement was the apprehension at the time caused by the structural adjustment programs of the IMF and World Bank, which tended to increase food insecurity for many, together with resistance to the WTO's Agreement on Agriculture. This was negotiated during the Uruguay Round of the General Agreement on Tariffs and Trade (GATT), which came into effect with the establishment of the WTO on January 1, 1995, and which imposed multilateral discipline on domestic agricultural policies.

### 6.8.2 Food Sovereignty and Food Security

Since 2008, the issue of food sovereignty as affecting food security has become hotly debated. The 2011–12 drought and famine experienced in the Horn of Africa, for instance, in which up to a quarter of a million people died, has not only underscored the need for countries to reflect on food security, but also prompted national policy shifts toward domestic food self-sufficiency.

In August 2008 when food prices reached their highest level in years, it became clear that no country could continue to rely on the market for its food needs. This applies even to rich countries, for when world stocks are very low, for whatever reason, their money may not guarantee that they receive food. This has ushered in a rethinking over whether countries, poor or rich, should rely as much as they have done in the past on the international marketing and trading system to ensure food security, or rather they better invest in domestic food production systems. Many African governments now believe that because markets can fail at any time, it is no longer possible to guarantee food security for their people in a business-as-usual manner.

Many public policy-makers now believe that growing more of the country's essential food needs is a precondition for genuine food security. Moreover, that the need for this is closely related to the issue of social justice and the rights of farmers and indigenous communities to better control their own future without being subject to the whims of the world market. That opens the question of who produces the food and where, and who controls the food market. The powerlessness of smallholders in Africa to have a say in whether their land is allocated to "Concessions" by the government as part of a national policy goal, and to prevent the dumping of cheap foreign food on the local market which can destroy their agricultural livelihoods, is being acknowledged.

In November 2012 the Committee on World Food Security (CFS) initiated a process to develop a set of principles to promote responsible investments in agriculture and food systems that contribute to food security and nutrition. The principles are intended for all stakeholders who are involved in, benefit from, or are affected by agricultural investments. A *Zero Draft* of the *Principles for Responsible Agricultural Investment (RAI)* was prepared in 2013. The RAI principles are intended to promote investments in agriculture that will enhance the food and nutrition security of people, especially those most exposed to the risk of hunger and undernutrition.

The CFS held a two-day regional consultation in Johannesburg, South Africa, in November 2013 on this Zero Draft, in the context of food and nutrition security in Africa. The consultation was organized to receive feedback from a broad range of stakeholders from the continent not only to improve the existing draft but also to foster ownership of the RAI principles within Africa. Stakeholders came from the public and private sectors, UN bodies and the African Union, and civil society—including farmers' organizations, the Forum for Agricultural Research in Africa (FARA) and NEPAD's Planning and Coordinating Agency (NPCA). Similar regional consultations were planned in Asia and the Pacific, Latin America and the Caribbean, Europe, the Near East and North America for early 2014, to be followed by a global meeting in Rome in May 2014, to negotiate the final version of the RAI document. *The Africa regional consultation flagged the issue of food sovereignty for discussion in Rome, and proposed that Africa's food policies should aim beyond assuring food security.*

The consultation in Johannesburg recognized that *there is an ongoing and unresolved debate about the merits of aiming for food sovereignty, noting that much of the main shortcoming of advocacy for food sovereignty is the absence of a robust base of evidence to lead it out of the purely rhetorical arena.* The consultation therefore recommended the establishment of the evidence required to support Africa's need for food sovereignty as a policy goal. FARA, in its capacity as the continental organization mandated to coordinate agricultural research in Africa, was requested to coordinate the establishment of the necessary evidence that would inform Africa's position for the negotiations in Rome in May 2014. FARA accepted to undertake this task ahead of the May meeting and fielded a consultancy mission entitled "Food sovereignty versus food security: where does Africa stand in the post-Millennium Development Goals era?".

The scale and quality of agricultural investments undertaken by African countries will depend largely on the path through which countries choose to achieve food and nutrition security. Therefore, achieving consensus on issues bordering on food security and food sovereignty is of vital importance to the adoption of the CFS Principles for Responsible Agricultural Investments in the context of food security and nutrition. It is noted that the term "food sovereignty" has no mention in the first draft of the CFS Principles for Responsible Agricultural Investments.

### 6.8.3 Evaluation of the Two Narratives

There are thus two ongoing narratives. The narrative on the one hand is espoused by food sovereignty (in the widest sense) advocates, who highlight a number of things in smallholder agricultural communities in developing countries which they believe, are fundamentally flawed. These include modern technologies which they view as harmful to the rural idyll (such as pollution using agrochemicals and GM crops), dependency on foreign markets which are seen as exploitative and undermine the civil rights of villagers, the concept of the commercial private sector and profit-driven enterprise viewed as being morally repugnant, and so on.

The opposing narrative points out that if all "improved technologies" used in the world's agriculture were to be abandoned overnight, for sure the world would starve in very short shrift. It is surely lack of such technologies which is responsible for corn yield in Mexico being only 57% of its potential on rainfed land of small-medium scale farm holdings, which leads to a one-third shortfall of its need and a food import bill of over $US20 billion per year (Fernández et al., 2013), representing a profound loss of its food self-sufficiency (sovereignty) over recent decades. The narrative maintains that being an impoverished smallholder is not a vocation of choice but one of default and lack of realized opportunity, through real-life opportunities not being organized and harnessed.

Such benefits do not come from inefficient farming under a subsidized State farming regime. Those opposed to many of the positions of the food sovereignty movement point to the socialist or communist utopian ideals of many a country in pursuit of food self-sufficiency, in both past and present times, in which most famines have occurred. For instance, Stalinist Russia after World War I, Sudan under President Nimeiry who overtaxed Gezira irrigation scheme farmers to extinction, Ethiopia and Cuba (both now changing), and North Korea. Nor has compulsory acquisition by government of large-scale private farms helped Zimbabwe, in which a national food surplus has been converted into food deficit and hyper-inflation, its initial food security transformed into food insecurity and poverty for the majority. In the 2014 election in South Africa, the political party which advocated the dissolution of efficient (white-owned) farms came nowhere near winning the national mandate. The painful stories of suffering as a result of such a policy from many Zimbabwean refugees seeking a better life in South Africa had been taken to heart.

The apparent competing narratives are thrown into the limelight sometimes in high-profile public disagreements. In one instance FAO, with its mandate of feeding the world, took central stage. It became deeply enmeshed in the complex web of politics, vested interests and passions linked with agricultural practice, when it drew up for the Nicaraguan parliament a draft law on national food security that contradicted the work on national food sovereignty that Nicaraguan civil society and policy-makers had accomplished over several years. The European Union Food Facility program being implemented by FAO was caught up in the confrontation (Müller, 2013).

Aerni (2011) *op. cit.* contends that Africa has become a net importer of food because political ideology has triumphed over knowledge gained from farmers' experience in the field and agricultural research. Food dumping of surpluses from the developed world into the developing world is part of this problem, accentuated by governments in the latter who have overtaxed farmers, subsidized consumer prices in the politically important urban environment and crowded-out private sector investment in agriculture. These policies undermine food security and need to be rethought.

Huge mistakes have been made on both the capitalist side and the authoritarian nationalist side. Now is the time for reconciliation and accommodation, a time to learn and agree actions based on evidence and exploiting a middle way which rewards judicious and calibrated investment and hard work. Paradoxically, it is the nominally communist regime of China, where one may find excellent pragmatic approaches to smallholder agriculture, in which using modern technologies and making a profit have been encouraged by the government, which deserves much credit for delivering the transformation of rural communities from grinding poverty. China's 200 million smallholders are now able to feed a population of 1.3 billion, and the country's poverty incidence decreased from 31% in 1978 to just 2.8% in 2008 (OECD, 2010). Governments of sub-Saharan Africa would do well to look to China as a template for what is possible to uplift their vast rural constituencies. Some excellent examples of this farmer-first inclusive pragmatic policy already exist in Africa, in which existing improved technologies are provided to farmers on a demand-led basis—the NGO-led One Acre Fund in Kenya and Rwanda shows that community development and market development are not antagonistic paradigms at all, as many "food sovereignty" advocates in affluent countries would have the world believe, but worthy bed-fellows in the real world (Juma, 2011).

The concerns of politically motivated food sovereigntists can be allayed on the ground through better-organized farming communities, which achieve control of their own destinies through pooling expertise and ideas, benefitting from economies of scale and greater involvement in marketing their produce. Through focusing on value-chains, exploiting the potential for adding value themselves to primary produce, through turning some of their milk into cheese and yoghurts, and attention to better handling, faster transport of olives to the oil mill (say), better cleanliness to prevent contamination to enable inspectors to give high quality status to products, more attractive packaging, more aggressive marketing, etc. None of this will happen in subsistence-level production and marketing.

So, in the debate between advocates and detractors of the total package of food sovereignty, there seems plenty of middle ground to be occupied, in which both sets of advocates work together to achieve common goals. This is the way to avoid the "unacceptable face of capitalism"[17] as seen in Liberia in 2011, for instance, as a result of price-fixing by local cartels, for rice and other essential commodities, causing hardship and hunger for poor people. Simultaneously, both advocates and detractors could work for the identification and use of improved crop varieties, as determined by the farmers (see Section 5.3.1.3), and other means to increase productivity in a sustainable way. There is no need for the poor to be exploited and impoverished by "the market"; rather the market is a tool to release smallholders from drudgery, the smallholders becoming part of the emerging commercial private sector.

Unfairness in the market can be countered by lobby groups, and developing and enforcing appropriate strategies, laws and regulations by an enlightened government and legislature. The international market in food is here to stay, though there is certainly scope for it to work better and more equitably. This is not just wishful thinking—who would have contemplated that the once all-powerful banks would have been so humbled over recent years by concerted pressure from the public and governments?

---

[17] A term used by former British Prime Minister Edward Heath in 1973.

Good evidence that this convergence of disparate views is indeed happening is to be noted in many of the papers presented at the Food Sovereignty Conference sponsored by the Program in Agrarian Studies at Yale University, in September 2013,[18] and the law on a productive community-based agricultural revolution promulgated in July 2011 in Bolivia. This law seeks to combine modern scientific farming techniques with ancestral indigenous traditions, and may have struck about the right balance in terms of food sovereignty, as it has drawn criticism from both the environmental and export-oriented agribusiness lobbies !

## 6.9 CLIMATE CHANGE

> *Without action at the global level to address climate change, we will see farmers across Africa – and in many other parts of the world, including in America – forced to leave their land. The result will be mass migration, growing food shortages, loss of social cohesion and even political instability.*
>
> **Kofi Annan, former UN Secretary-General (1997–2006) (opening the Center on Food Security and the Environment, Stanford University, USA, November 2011) (http://kofiannanfoundation.org/newsroom/speeches/2011/11/food-security-global-challenge)**

### 6.9.1 Introduction

Climate change brought on by human activities is having a major impact worldwide. Temperatures are rising (World Bank, 2012), rainfall patterns changing, polar ice caps melting, glaciers shrinking, permafrost thawing, sea level rising, and frequency and depth of extreme weather events are increasing. Climate is becoming more unstable and less predictable.

The World Bank predicts temperature increases of 4°C by the end of the 21st century. According to the Intergovernmental Panel on Climate Change (IPCC), most of the observed increase in global average temperatures since the mid-20th century is likely due to the observed increase in manmade "greenhouse gas" (GHG) concentrations. (GHG emissions are mainly carbon dioxide, but also methane, nitrous oxide and fluorinated gases).[19]

Direct GHG emissions from agriculture include methane emissions from livestock and flooded rice fields, nitrous oxide emissions from the use of fertilizers, and carbon dioxide emissions from loss of soil organic carbon as a result of certain agricultural practices. Agriculture is also responsible for emissions from other sectors (agrochemical industry, farm tillage and produce transport, and energy supply for farm and value-adding processing).

[18] Retrieved from http://www.yale.edu/agrarianstudies/foodsovereignty/abstracts.html (accessed March 10, 2015).

[19] Retrieved from www.epa.gov/climatechange/ghgemissions/climatechange (United States Environmental Protection Agency—EPA)

Agricultural activities, including indirect effects through deforestation and other forms of land conversion which reduce the absorption of carbon dioxide by green plants, account for about a third of total global warming potential from current GHG emissions. Therefore, attempts to reduce the direct and indirect emissions from agriculture represent an important ingredient of overall effort to slow the pace of climate change.

While global predictions are more easily testable, climate models are insufficiently precise to predict impact at the local and regional scales. It is clear however, that each one of us without exception is being, and will be, affected ... not just a few remote islanders or polar bears (Elver, 2014)! Climate change is one of the greatest potential obstacles to ending poverty and one of the gravest equity challenges of our time. While rich countries have been historically most responsible for causing climate change, it is poor countries which are being hit hardest by its effects. There is need for concerted individual and communal targeted actions to address the threat cited in Senator Al Gore's film "An inconvenient truth", which in 2006 introduced the accumulating bad news of climate change to the then largely unaware public domain. Shortly after that, in 2008, FAO published a similar strong advisory on how we humans are affecting climate change, with dire consequences (quoted in Box 6.6).

### Box 6.6 Climate Change and Food Security

*Climate change will affect all four dimensions of food security: food availability, food accessibility, food utilization and food systems stability. It will have an impact on human health, livelihood assets, food production and distribution channels, as well as changing purchasing power and market flows. Its impacts will be both short term, resulting from more frequent and more intense extreme weather events, and long term, caused by changing temperatures and precipitation patterns.*

*People who are already vulnerable and food insecure are likely to be the first affected. Agriculture-based livelihood systems that are already vulnerable to food insecurity face immediate risk of increased crop failure, new patterns of pests and diseases, lack of appropriate seeds and planting material, and loss of livestock. People living on the coasts and floodplains and in mountains, drylands and the Arctic are most at risk.*

*As an indirect effect, low-income people everywhere, but particularly in urban areas, will be at risk of food insecurity owing to loss of assets and lack of adequate insurance coverage. This may also lead to shifting vulnerabilities in both developing and developed countries.*

*Food systems will also be affected through possible internal and international migration, resource-based conflicts and civil unrest triggered by climate change and its impacts.*

*Agriculture, forestry and fisheries will not only be* affected *by climate change, but also* contribute *to it through emitting greenhouse gases. They also hold part of the remedy, however; they can contribute to climate change mitigation through reducing greenhouse gas emissions by changing agricultural practices.*

*At the same time, it is necessary to strengthen the resilience of rural people and to help them cope with this additional threat to food security. Particularly in the agriculture sector, climate change adaptation can go hand-in-hand with mitigation. Climate change adaptation and mitigation measures need to be integrated into the overall development approaches and agenda....".*

**Source: *Excerpt from the Foreword of FAO (2008). Climate Change and Food Security: a framework document (107 pp).* ftp://ftp.fao.org/docrep/fao/010/k2595e/k2595e00.pdf *(reproduced with permission).***

Threats to food security abound as a result of this induced climate change, and are already taking their toll, reducing past gains in the fight against poverty. The wide-ranging threat to the Amazonia region of South America, for instance, is indicated in Case Study 5 on the book's companion website.

FAOSTAT's world food balance sheet data[20] show that some 99% of human food (calories) comes from the land, and only 1% from oceans and other aquatic systems. Maintaining and increasing the world's food supply relies above all on maintaining productivity of the soil year on year, and water supply is clearly one of the key factors involved. Both floods and drought contribute to the estimated 10 million hectares of crop land lost to soil erosion each year, the rate of loss being a factor of 10–40 times more than the rate of soil formation (Pimental and Burgess, 2013).

Not only does climate change have a *direct* effect on agricultural and fishery productivity, but *natural disasters* brought on by climate change have an enormous cost to human populations. Standing crops and food stocks can be washed away in floods or burnt by fire, for instance. Hurricane damage in Central America has devastating effects, as detailed for Belize in Case Study 3, and the intensity and frequency of these seem to be increasing. The floods in Pakistan in 2010 affected more than 20 million people and caused damage of over US$10 billion. Women are especially vulnerable in such disasters. For example, in Myanmar during Cyclone Nargis in 2008, 61% of those killed were women.

While such events in *developing* countries directly affect the people living there, climate-related damage in *developed* economies can also affect people in developing countries. For instance, droughts over the last decade in Russia and Australia, which have lowered their grain harvests, have contributed to higher *global* food prices, according to the micro-economics laws of supply and demand in a free market.

## 6.9.2 Effects From Melting Polar Glaciers and Ice Caps

### 6.9.2.1 COASTAL

A study team of glaciologists led by the US National Aeronautics and Space Administration (NASA) has reported that key glaciers in West Antarctica are in irreversible retreat, though their total disappearance may take centuries.[21] The basis of this conclusion was interpretation of actual observations over 40 years of six ice streams draining into Amundsen Bay, which would raise global sea level by 1.2 m (4 ft). Their retreat and collapse will also influence adjacent sectors of the West Atlantic Ice Sheet, which could triple this contribution to sea level. The study team attributed these observations to the

---

[20] Retrieved from www.faostat.fao.org/site/368/default.aspx#ancor (accessed May 4, 2014). The figure of 1% is likely to be a significant underestimate, however, due to the "invisibility" to statisticians of much of the artisanal fishery subsector (see Section 5.3.1.7).

[21] Retrieved from www.bbc.com/news/science-environment-27381010 (accessed May 12, 2014). "Nothing can stop retreat" of West Antarctic glaciers.

drivers of global warming and loss of ozone in the stratosphere. The 1.2 m mentioned is still a small fraction of the total sea level rise potential of a melting Antarctic ice cap, which holds around 26.54 $km^3$ of ice [or 58 m (190 + ft) of sea level rise equivalent] as identified by the ongoing *Bedmap2* project. The changes that are occurring, and the speed with which they are happening, are continually monitored.

Melting glaciers at the poles pose existential threats to whole countries, such as The Maldives and Bangladesh. Only the speed of the inundation is in doubt, not the result. As polar ice melts, it affects people unevenly—it is the poor who are affected first and worst. Yet not only the poorest and in developing countries—a 10-ft rise in sea level, which is far less than that predicted from just the Antarctic glaciers referred to above, would put the whole of Miami under water, for instance. Fig. 6.2 shows the threat to many cities around the coast of the African continent, both their slums and their smart districts. Not only are the living being affected, but even the dead.[22]

#### 6.9.2.2 INLAND

Ice melt is affecting inland centers of population too. In the high mountains of the Andes, the Himalayas, Hindu Kush and elsewhere, the slow spring melt of glaciers provides water for agricultural areas in the foothills over the agricultural year, and for domestic and industrial use in the towns and cities lower down. Glaciers have provided a store of such dry season agricultural water since settled agriculture began, to supplement current rainfall and that released gradually as rain percolates through forest and bog ecosystems higher on the slopes. Yet this glacial bounty is in prospect of coming to an end. These glaciers have been melting more quickly as a result of global warming, evidence for this having been seen for twenty years now, as reported by the World Glacier Monitoring Service. Not only is ice in lower parts of the glaciers melting increasingly early in the year, but the leading edge has been retreating ever further up the slope.

Glacier melt characteristics serve as early indicators of climate change (Thompson, 2010). Thompson recorded his research over the previous 35 years, on ice-core records of climatic and environmental variations from polar regions and low-latitude high-elevation ice fields of 16 countries. The ongoing widespread melting of high-elevation glaciers and ice caps, particularly in low-to-middle latitudes, provides some of the strongest evidence to date that a large-scale pervasive, and, in some cases, rapid change in Earth's climate system is under way. His paper highlights observations of 20th and 21st century glacier shrinkage in the Andes, the Himalayas and on Mount Kilimanjaro. The current warming is unusual when viewed from a millennial perspective provided by multiple lines of proxy evidence, and the 160-year

[22] Retrieved from www.bbc.com/news/science-environment-27742957 (accessed June 7, 2014). "Climate change helps seas disturb Japanese war dead".

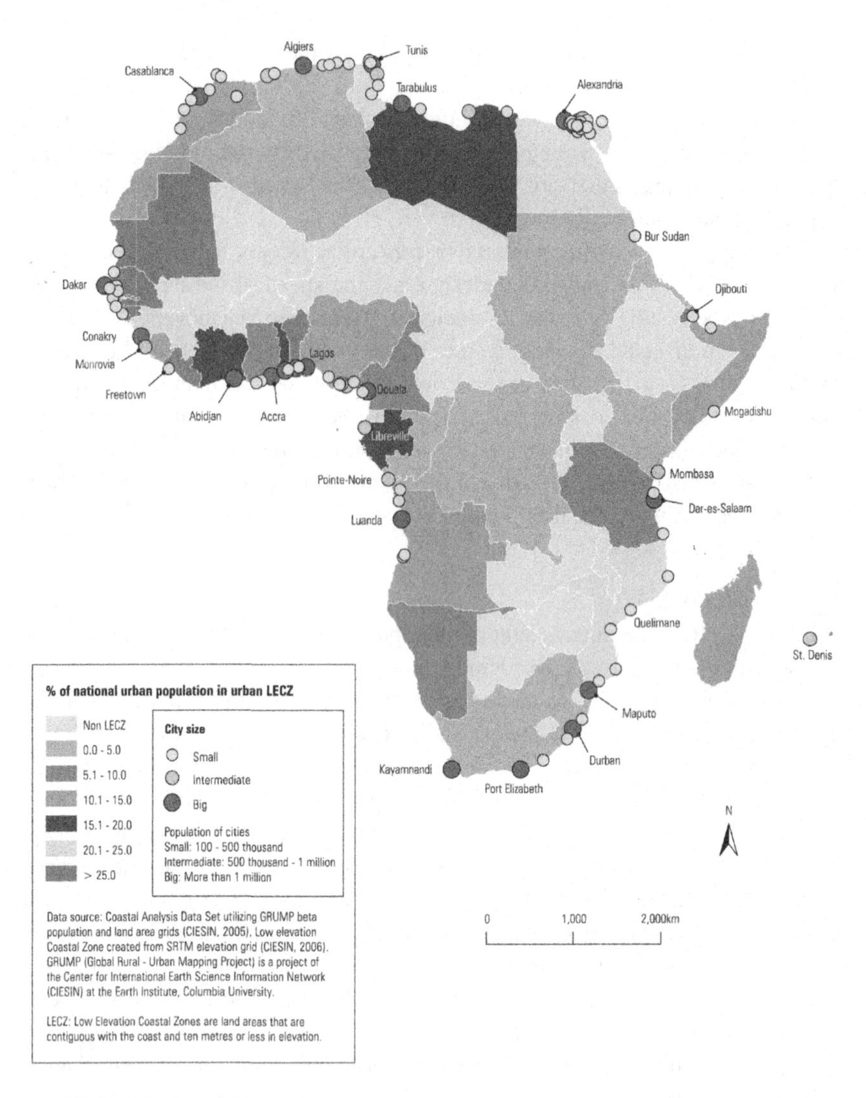


**FIGURE 6.2** African cities at risk due to sea level rise. Reproduced with permission from UN-Habitat (2014).

record of direct temperature measurements. Despite all this evidence, plus the well-documented continual increase in atmospheric greenhouse gas concentrations, the planet's population has corporately taken little action to address the global-scale problem. Hence, the rate of global carbon dioxide emissions continues to accelerate, and global warming gathers pace. As a result of our inaction, we have three options Thompson says: mitigation, adaptation or suffering !

Between 1975 and 2006, the glacier area in the Bolivian Andes shrank by almost a half, and its once-famous Chacaltaya glacier has melted completely.

Much of the glacial coverage of the Peruvian Andes has disappeared, and the flow of water to the country's arid coastal region, in which 60% of the country's population lives, and in which the capital Lima is situated, will experience ever-reducing river water flow. A UN study refers to Lima as "a crisis waiting to happen". Hydropower supply will dwindle and irrigation water for potato and wheat crops will decline. Food, water, energy and political securities are all thereby compromised, all because of global warming.

In another study (Wiltshire, 2013), through examining the climate drivers of change rather than the glaciological response, in the Hindu-Kush which separates central from southern Asia, Wiltshire has determined that the eastern glaciers are projected to decline over the 21st century despite increasing precipitation, with the western glaciers expected to decline at a slower rate. Overall, the eastern Himalayan glaciers are expected to be most sensitive to climate change due to the decreases in snowfall and increased "ablation" associated with warming. Just as for the Andean region, water, food and energy securities of many countries in Asia are thereby threatened. Box 6.7 contains a quote from celebrated environmental analyst Lester R. Brown, pointing out the huge threat posed by ice melting in the Himalayas and on the Tibetan Plateau (Brown, 2011).

Lesser order environmental catastrophes are affecting Africa too. In the Rwenzori mountain range on the equator, on the border of Uganda and DR Congo, there has been a massive shrinkage and even disappearance of some named glaciers over the last century (Kaser, 2002). The sustainable agricultural livelihoods of the Bakonjo people of the Rwenzori foothills have depended on the melt water of these glaciers, as have the remarkable ecosystems which constitute the UNESCO World Heritage site, the Rwenzori National Park. A similar retreat has been recorded of the glaciers on Mount Kilimanjaro in Tanzania, with the ice cap predicted to reduce by two thirds or even disappear completely by 2060. This will progressively negatively affect the hitherto-sustainable irrigated agricultural livelihoods of the Chagga people on the lower slopes.

### 6.9.3 The IPCC and Potential Climate Change Adaptation/Mitigating Measures

The IPCC is an intergovernmental body, with 195 member countries, the mandate of which is the assessment of climate change.[23] It was established by UNEP and the World Meteorological Organization (WMO) in 1988, to provide a clear scientific view on the current state of knowledge on climate change, and its potential environmental and socioeconomic impacts. In the same year, the UN General Assembly endorsed the IPCC. Thousands of scientists from all over the world contribute to the work of the IPCC on a

[23] Retrieved from www.ipcc.ch/organization/organization.shtml (accessed May 3, 2014).

## Box 6.7 Impact of Inland Glacier Melt for China and India

*......Ice melting in the Himalayas and on the Tibetan Plateau poses an even graver threat to food security at a global scale. It is the ice melt from these mountain glaciers that helps sustain the major rivers of Asia during the dry season, when irrigation needs are greatest. In the Indus, Ganges, Yellow, and Yangtze River basins, where irrigated agriculture depends heavily on the rivers, the loss of any dry-season flow is bad news for farmers. China is the world's leading producer of wheat. India is number two. With rice, China and India totally dominate the world harvest. Therefore, the melting of these glaciers coupled with the depletion of aquifers present the most massive threat to food security the world has ever faced. In India, the giant Gangotri Glacier, which helps keep the Ganges River flowing during the dry season, is retreating. The Ganges River is by far the largest source of surface water irrigation in India and a source of water for the 407 million people living in the Gangetic basin.*

*Yao Tandong, a leading Chinese glaciologist, reports that glaciers on the Tibetan Plateau in western China are now melting at an accelerating rate. Many smaller glaciers have already disappeared. Yao believes that two thirds of these glaciers could be gone by 2060. If this melting of glaciers continues, Yao says it "will eventually lead to an ecological catastrophe". The Yangtze, by far the country's largest river, helps to produce half or more of its 130-million-ton rice harvest (per year).*

*Like the depletion of aquifers, the melting of glaciers can artificially inflate food production for a short period. At some point, however, as the glaciers shrink and the smaller ones disappear entirely, so does the water available for irrigation. The melting of the glaciers on the Tibetan Plateau would appear to be China's problem. It is. But it is also everyone else's problem. In a world where grain prices have recently climbed to record highs, any disruption of the wheat or rice harvests due to water shortages in India or China will raise their grain imports, driving up food prices for all.*

*In India, where just over 40 percent of all children under five years of age are underweight and undernourished, hunger will intensify and child mortality will likely climb. For China, a country already struggling to contain food price inflation, there may well be spreading social unrest if food supplies tighten. For U.S. consumers, this melting poses a nightmare scenario. If China enters the world market for massive quantities of grain, as it has already done for soybeans over the last decade, it will necessarily come to the United States—far and away the leading grain exporter.*

*Ironically, the two countries that are planning to build most of the new coal-fired power plants, China and India, are precisely the ones whose food security is most massively threatened by the carbon emitted from burning coal. It is now in their interest to try and save their mountain glaciers by quickly shifting energy investment from coal-fired power plants into energy efficiency, wind farms, and solar thermal and geothermal power plants.*

**Source: *Retrieved from* Brown (2011). *(Reproduced with permission of the Earth Policy Institute, Washington DC, USA.)***

voluntary basis. Based in Geneva, IPCC reviews and assesses the most recent scientific, technical and socio-economic information produced worldwide relevant to the understanding of climate change. It does not conduct any research itself. The IPCC provides rigorous and balanced scientific information to decision-makers. By endorsing the IPCC reports, governments acknowledge the authority of their scientific content. The work of the organization is therefore policy-relevant and yet policy-neutral, never policy-prescriptive. The IPCC and US environmental activist and politician Al Gore were jointly and equally awarded the 2007 Nobel Peace Prize.

The 5th assessment report of the IPCC was released on April 1, 2014. It comprises three working group contributions,[24] the synthesis report due to be formulated during 2014 and tabled for consideration in Copenhagen at the end of October, 2014. The report provides an irrefutable scientific consensus on how the rising level of "greenhouse gases" from human activities impacts agricultural and aquatic food production systems. Climate change will worsen problems that society already has, such as poverty, sickness, violence and refugees, according to the report. Lower wheat, rice and corn (maize) yields, and diminished fish catches are also cited, and increased levels of food insecurity among the world's poorest people. The report concludes *inter alia* that climate change is already reducing food production and driving up food prices, and will continue to exert its harmful influence on food security, especially in the tropics, with the prospect of this effect intensifying in the decades to come. However, food production at higher latitudes, such as northern China and Europe, could in some instances benefit from climate change.

The report predicts that unless mitigating measures are taken, crop yields in Africa and southern Asia could decline by 8% by 2050, and yields of tropical fisheries by as much as 40%. To make matters worse, the human population dependent on the decreasing production is estimated to rise, indeed doubling by then what it is now in Africa, for instance. The report indicates that as a result of climate change, *crop yields could decline by 2%* per *decade*, while *demand is increasing by 2%* per *year*.

A huge cost tag can be apportioned to climate change. For instance, at the Rio + 20 Earth Summit in Brazil in 2012, a study was released showing that climate change will cost Latin America $US100 billion per year by 2050 (Vergara et al., 2013). That covers falling harvests, damage caused by natural disasters such as flooding, drought and the spread of tropical diseases to new areas, and $7 billion from lost fishery and tourism revenues as a result of coral bleaching in the Caribbean, as warmer waters kill the coral reefs. And these calculations are based on a global temperature rise of just 2°C, which is the "best-case scenario", set by the United Nations Framework Convention on Climate Change (UNFCCC) negotiations in Cancun, Mexico in 2010. However, the cost of investments in adaptation to address these impacts is much smaller, in the order of one tenth of the estimated physical damages, according to the study.

On April 3, 2014, a meeting convened in London, entitled "Agriculture growth, jobs, food security and climate: taking action in response to IPCC", focused on the implications of the 5th IPCC assessment report, and what actions the world must take to help farmers adapt to climate change in the interests of assuring a food-secure and prosperous world. It was organized by the CGIAR Research Program on Climate Change, Agriculture and Food Security (CCAFS), IFAD, the World Bank, DfID and the International Sustainability Unit, and global risk adviser Willis Group Holdings Plc. Delegates included

[24] Climate Change 2013: The physical science basis; Climate Change 2014: Impacts, adaptation and vulnerability; Climate change 2014: Mitigation of climate change.

representatives from the research community, farmers' organizations and other civil society groups, the private sector and donor agencies.

In addition to pointing out the inescapable imperative for action by governments to stem greenhouse gas production, the London meeting reinforced the options available to mitigate the effects of global warning. These include breeding and selection for drought-tolerant crop varieties. Yet development of such varieties, testing and distributing them, together with other adaptive innovations to improve resilience and sustainability of agricultural livelihoods, require strong investment in smallholder farming and pastoralism *now*. Many of these varieties can be developed regionally and spread through the CGIAR system, such as by ICARDA for dryland west Asia, and wider use of the on-farm research-extension protocol called Participatory Variety Selection (PVS)(see Section 5.3.1.3). These innovations need to target the 500 million smallholdings in the developing world which together produce 80% of food production there. The evidence to support such investment has clearly been provided by the 5th IPCC assessment report.

Other technical innovations currently being implemented by many stakeholder groups serving to adapt the developing world's productive base to climate change, include the minimum tillage Conservation Agriculture technique pioneered in the United States and finding resonance with smallholders in Zambia, for instance (see Case Study 2 on the book's companion website). Risk insurance to compensate livestock herders for drought losses in eastern Africa is being pioneered by DfID and the International Livestock Research Institute, and desert reclamation work is being spearheaded by IFAD in Nigeria and Niger. There is also huge scope for rolling out to other countries the hugely successful Productive Safety Net Program (PSNP) in Ethiopia, as is currently being done in Malawi, Tanzania and Rwanda (see Case Study 1). The program Building Resilience and Adaptation to Climate Extremes and Disasters (BRACED) is helping build resilience, especially of women to disasters, DFID providing grants to 21 projects led by NGOs to scale up activities in South Asia and Africa with a focus on improving agricultural practices.

Adaptation mechanisms need to look beyond the farm gate too, at other facets of food security which are affected by climate change, such as crop storage, transport networks and markets. The private sector has an important role to play in building resilience to climate change, and some initiatives can take the form of Public-Private Sector ventures. At the World Economic Forum in Davos, Switzerland, in January 2014, the global business community identified water supply for agriculture and industrial processes as one of the top four risks confronting it.

### 6.9.4 Reducing the Speed and Intensity of Climate Change

Yet mitigating the effects of climate change comprises a rearguard action, to combat change which has already happened or is in the pipeline.

The overarching challenge is to *reduce* greenhouse gas emissions which accelerate the rate at which such climate change will proceed in future. World leaders need to work hard for binding agreements which reduce greenhouse gas emissions and keep increases in the global average temperatures to under 2°C, in the first instance. The 5th IPCC assessment report of 2014 *op. cit.* states that warming of over 4°C, a likely outcome of continued business-as-usual practices, will cause severe, pervasive and irreversible change. This must surely be avoided at all costs.

Since the 1990s there have been attempts to devise legally binding agreements to limit greenhouse gas emissions. These have been largely unsuccessful. Representing the world's large energy companies, the biggest contributors to greenhouse gas emissions, the powerful fossil fuel lobby in the United States empowers legislators to diminish or deny the effects of climate change largely attributed to burning of fuel. Hillel Elver, in her *Al Jazeera* article of 2014 *op. cit.*, said that after the 5th IPCC Assessment Report was published, some US Congressmen proposed a bill seeking to limit the research activities of the National Oceanic and Atmospheric Administration to exclude climate change from its mandate and focus only on forecasting severe weather events ! Yet it will be the United States which stands to lose more than many by continuing this charade and not standing up to these companies and lobbyists. With an under-performing wheat belt under permanent drought one day perhaps, it will be US citizens paying exorbitant prices for their bread, made using imported wheat, if available supplies can be found. Meanwhile, many cities in Latin America, for instance, are far ahead of their US counterparts in preparing contingency plans for climate change, according to a study by the Massachusetts Institute of Technology (MIT, 2012)—Quito, for example, is investing heavily to maintain its water supply as glaciers melt above the Ecuadorean capital.

As part of the Kyoto protocol on greenhouse gas emissions, adopted in 1997 which came into force in 2005, 37 industrialized countries (including the 28 countries of the European Community) committed to reduce GHG emissions to an average of 5% against 1990 levels, during the first commitment period (2008–12). During the second commitment period (planned for 2013–20), parties committed to a legally binding reduction in GHG emissions by at least 18% below 1990 levels. However, the composition of parties in the second commitment period was different from the first. In addition, for the second period, over 70 other countries—both developed and developing—have made different types of non-binding commitments to reduce, or limit the growth in, their greenhouse gas emissions. The United States is not a signatory to the Kyoto protocol, nor is China or India.

Following an initiative of the European Union and the most vulnerable developing nations tabled at the Durban climate conference in December 2011, negotiations are under way to develop a new international climate change agreement that will cover all countries, through a process known as the Durban Platform for Enhanced Action. This agreement needs to bring together the current patchwork of binding and non-binding arrangements under the UN climate convention into a single comprehensive regime. The 21st session of the Conference of the Parties to the UNFCCC, scheduled for December 2015, in Paris, France, is charged with brokering and adopting a global deal to curb climate change, to be implemented from 2020. It will take the form of a protocol, another legal instrument or "an agreed outcome with legal force", and will be applicable to all Parties. Achieving a deal in Paris in 2015 will be a milestone toward a rational solution to the prospect of climate change catastrophe. This meeting will take place as this book goes to press.

### 6.9.5 Differential Vulnerability to Climate Change—Induced Food Insecurity

A Report by the CFS in 2012 (HLPE, 2012) *op. cit.* underlines the huge challenges which climate change superimposes on pre-existing food security vulnerability, reducing the productivity of most existing food systems and harming the livelihoods of those already at risk of food insecurity. Dryland agriculture in arid and semi-arid regions, where over 40% of the world's population live, including more than 650 million of its poorest and most food-insecure, is particularly vulnerable to drought and increased risk of food insecurity. Also, in many coastal regions there are large agricultural areas which are at risk of increased salinity of ground water and flooding, associated with rise in sea level. Climate change modeling suggests that the tropics will be the most-affected by climate change. Many of the poorest countries occur there, and may be the least able to adapt, and hence require the most help to do so.

Research suggests that extreme weather events like El Niño will become more intense as global temperatures rise. El Niño-Southern Oscillation (ENSO) relates to a warming of the Pacific Ocean, part of a complex cycle linking ocean and atmosphere and known to disrupt weather patterns around the world. In May 2015 the Australian Bureau of Meteorology warned that another strong El Niño effect was taking shape for the year, with a 70% chance of occurrence. The last strong El Niño five years previously was associated with poor monsoons in Southeast Asia, droughts in southern Australia, the Philippines and Ecuador, heatwaves in Brazil, extreme flooding in Mexico and blizzards in the United States. In August 2015 the Panama Canal Authority declared that it was "temporarily" reducing the size (draft) of ships allowed through the canal because of the drought associated with El Niño, which had lowered water levels in the component Gatun and Alhajuela lakes. Some 20% of current traffic may be affected, meaning that transit fee revenues will be reduced from the $1 billion present annual level, this likely to have a domino effect on the food security of Panamanian citizens.

Food security vulnerability to climate change starts at farm level, with the response of crops and livestock to water and temperature environmental stresses. These effects can reduce productivity in rural areas directly, while local and international markets transfer the effects to urban areas and beyond.

A more resilient food system can be brought about through technical efficiencies to irrigation systems or introduction of more drought- and/or saline-tolerant varieties.

One of the important recommendations in the HLPE (2012) report (in its Section 5.1) is that climate adaptation policies and programs should be complementary to, not independent of, those fostering food security. Climate change is but one of a range of threats to food security. Interventions designed to increase resilience of food systems are very likely to also contribute to climate change adaptation. In so doing, farmers should be at the center of these efforts, and location-specific approaches be devised ensuring that community needs are met, based on local knowledge, skills and other assets (see Case Study 9 on the book's companion website, for a discussion of how climate change can affect food prices).

Climate change disproportionately impacts women, in part because traditional domestic responsibilities usually fall on women and girls. Women are the primary caretakers of families and main managers of everything from food production to water management in their households. They also cook and clean, provide healthcare and hygiene. While being the main victims of climate change, their personal and intimate experience of its harsh impacts gives them keen insights into solutions to combat the problem. Yet rural women's perspectives are not at the forefront of the ongoing climate change debate; this must change too.

## REFERENCES

ACQUIRE Project, 2005. Moving Family Planning Programs Forward: Learning from Success in Zambia, Malawi, and Ghana. New York, NY, USA: The ACQUIRE Project/EngenderHealth. USAID.

Aerni, P., 2011. Food sovereignty and its discontents. ATDF J. 8 (1/2), 23–39.

Asian Development Bank, 2013. Gender Equality and Food Security—Women's Empowerment as a Tool Against Hunger. ADB with FAO, Mandaluyong City, Philippines, 114 pp.

BBC News, 2015. Anti-poverty pioneer wins 2015 World Food Prize, by Mark Kinver, July 2, 2015. Retrieved from: <http://www.bbc.co.uk/news/science-environment-33283680> (accessed 02.07.15.).

Brown, L.R., 2011. Rising Temperatures Melting Away Global Food Security. July 6, 2011. Retrieved from: <www.treehugger.com/clean-technology/rising-temperatures-melting-away-global-food-security.html> (accessed on 03.05.14.).

Burnett, K., Murphy, S., 2014. What place for international trade in food sovereignty? J. Peasant Stud. 41 (6), 1065–1084, Oxford: Routledge.

de Janvry, A., Sadoulet, E., 2001. Access to land and land policy reforms. In: de Janvry, A., Gordillo, G., Sadoulet, E., Platteau, J.-P. (Eds.), Access to Land, Rural Poverty and Public Action. Oxford Scholarship Online. October 2011. Available from: http://dx.doi.org/10.1093/acprof:oso/9780199242177.003.0001.

Elver, H., 2014. Climate Change and the Food Security Dimension. April 25, 2014. <www.aljazeera.com/indepth/opinion/2014/04/climate-change-food-security-di-201442472225837148>.

FAO, 2011a. The State of the World's Land and Water Resources for Food and Agriculture (SOLAW)—Managing Systems at Risk. FAO Rome and Earthscan, London, 308 pp.

FAO, 2011b. Good food security governance: the crucial premise to the twin-track Approach: background paper. In: Agricultural Development Economics Division (ESA) of FAO. Workshop, Rome, 5–7 December 2011. 41 pp.

FAO, IFAD, WFP,2013. The state of food insecurity in the world 2013: the multiple dimensions of food security. Executive Summary. 4 pp. <http://www.fao.org/docrep/018/i3458e/i3458e.pdf>.

Fernández, A.T., Wise, T.A., Garvey, E., 2013. Achieving Mexico's maize potential. In: Food Sovereignty: A Critical Dialogue. International Conference, Yale University, USA, September 14–15, 2013. Conf. Paper #10. 45 pp.

HLPE, 2012. Social protection for food security. A Report by the High Level Panel of Experts on Food Security and Nutrition of the Committee on World Food Security, Rome 2012. HLPE Report 4, June 2012. 100 pp. Retrieved from: <www.fao.org/fileadmin/user_upload/hlpe/hlpe_documents/HLPE_Reports/HLPE-Report-4-Social_protection_for_food_security-June_2012.pdf> (accessed 28.05.15.).

IDS, 2014. Gender and food security: towards gender-just food and nutrition security. Bridge Development—Gender. Institute of Development Studies, Sussex, UK, 104 pp.

IDS, 2015. Gender and Food Security. In Brief. Bridge Development—Gender. Institute of Development Studies, Sussex, UK, 8 pp.

Juma, C., 2011. New Harvest: Agricultural Innovation in Africa. Oxford University Press, New York, NY.

Kappeler, A., 2013. The perils of peasant populism: why redistributive land reform and "food sovereignty" can't feed Venezuela. In: Food Sovereignty: A Critical Dialogue. International Conference, Yale University, USA, September 14–15, 2013. Conf. Paper #65. 24 pp.

Kaser, G., 2002. Modern glacier fluctuations on the Rwenzori. In: Kaser, G., Osmaston, H.A. (Eds.), Tropical Glaciers. International Hydrology Series. Cambridge University Press, Cambridge, UK, pp. 63–116 (Chapter 6).

Madsen, E.L., 2011. Rwanda: Dramatic Uptake in Contraceptive Use Spurs Unprecedented Fertility Decline. NewSecurityBeat Blog, November 8, 2011. <http://www.newsecuritybeat.org/2011/11/building-commitment-to-family-planning-rwanda>.

Ministry of Health, 2009. Annual Report 2008. Ministry of Health, Kigali, Rwanda.

MIT, June 4, 2012. Survey: Latin American and Asian Cities Lead Way in Planning for Global Warming. P. Dizikes. MIT News Office, Massachusetts Institute of Technology, USA. <http://mitei.mit.edu/news/survey-latin-american-and-asian-cities-lead-way-planning-global-warming>.

Muhoza, D.N., Rutayisiri, P.C., Umubyeyi, A., 2013. Measuring the success of family planning initiatives in Rwanda: a multivariate decomposition analysis. Demographic and Health Survey Working Paper, No 94. ICF International, MD, USA. February 2013. USAID, 32 pp.

Müller, B., 2013. The temptation of nitrogen: FAO guidance for food sovereignty in Nicaragua. In: Food Sovereignty: A Critical Dialogue. International Conference, Yale University, USA, September 14–15, 2013. Conf. Paper #33. 27 pp.

Mwaikambo, L., Speizer, I.S., Schurmann, A., Morgan, G., Fikree, F., 2011. What works in family planning interventions: a systematic review. Stud. Fam. Plan. 42 (2), 67–82.

National Institute of Statistics [Rwanda], ORC Macro, 2006. Rwanda Demographic and Health Survey 2005. National Institute of Statistics of Rwanda and ORC Macro, Calverton, MD, USA.

National Institute of Statistics of Rwanda (NISR) [Rwanda], Ministry of Health (MOH) [Rwanda], ICF International, 2012. Rwanda Demographic and Health Survey 2010. NISR, MOH and ICF International, Calverton, MD, USA.

OECD, 2010. Agricultural Transformation, Growth and Poverty Reduction—How it Happened in China, Helping it Happen in Africa. China-DAC study group, 28 pp (summary of discussions at an international conference on Agriculture, Food Security and Rural Development for growth and poverty reduction. Bamako, Mali on 27–28 April 2010).

Opitz-Stapleton, S., 2014. Understanding Gendered Climate Variability and Change Impacts in Jinping and Guangnan counties, Yunnan Province, China. 42 pp. <www.intasave.org.cn>.

Patterson, M., 2004. The Shiwa pastures, 1978–2003: Land tenure changes and conflict in northeastern Badakhshan. Afghanistan Research and Evaluation Unit (AREU) Case Studies Series. 55 pp.

Pimental, D., Burgess, M., 2013. Soil erosion threatens food production. Agriculture. 3, 443–463.

Prosterman, R.L., Hanstad, T., 2003. Land Reform in the twenty first century: new challenges, new responses. RDI Reports on Foreign Aid and Development # 117, Seattle, USA. 32 pp.

Schuler, S.R., Hashemi, S.M., Jenkins, A.H., 1995. Bangladesh's family planning success story: a gender perspective. Int. Fam. Plan. Perspect. 21 (4), 132–137&166.

Sisto, I., 2007. Gender key to food security and food safety. In: Sagardoy, J.A., et al., (Eds.), Mainstreaming Gender Dimensions in Water Management for Food Security and Food Safety. CIHEAM, Bari, pp. 17–25. (Options Méditerranéennes : Série A. Séminaires Méditerranéens; n. 77). 2. Regional Coordination Workshop of GEWAMED (Mainstreaming Gender Dimensions into Water Resources Development and Management in the Mediterranean Region), 2007/03/12-14, Larnaca (Cyprus) <http://om.ciheam.org/om/pdf/a77/00800475.pdf>.

Solo, J., 2008. Family Planning in Rwanda: How a Taboo Topic became Priority Number One? IntraHealth, Chapel Hill, NC, USA, 35 pp.

Thompson, L.G., 2010. Climate change: the evidence and our options. Behav. Anal. 33 (2), 153–170.

UNDP, 2012. Gender, agriculture and food security. Gender and Climate Change: Capacity Development Series. Africa. Training Module 4. 36 pp.

UN-Habitat, 2008. Secure Land Rights for All. Global Land Tool Network, Nairobi, Kenya, 47 pp.

UN-Habitat, 2014. The State of African Cities 2014: Re-imagining Sustainable Urban Transitions. 278 pp. <www.unhabitat.org/the-state-of-african-cities-2014/>.

Vergara, W., Rios, A.R., Galindo, L.M., Gutman, P., Isbell, P., Suding, P.H., et al., 2013. The Climate and Development Challenge for Latin America and the Caribbean: Options for Climate Resilient Low Carbon Development. Inter-American Development Bank, Economic Commission of Latin America & the Caribbean, World Wildlife Fund, <http://publications.iadb.org/handle/11319/456?locale-attribute = en>.

Wickeri, E., Kalhan, A., 2010. Land rights issues in international human rights law. Malays. J. Hum. Rights. 4 (10), 10, <http://www.ihrb.org/pdf/Land_Rights_Issues_in_International_HRL.pdf>.

Wiltshire, A.J., 2013. Climate change implications for the glaciers of the Hindu-Kush, Karakoram and Himalayan region. Cryosphere Discuss. 7 (4), 3717–3748. Available from: http://dx.doi.org/10.5194/tcd-7-3717-2013.

World Bank, 2009. Gender in Agriculture Sourcebook. Module 1: Gender and Food Security. World Bank, FAO and IFAD, pp. 11–22 (of 792 pages).

World Bank, 2010. Social Dimensions of Climate Change. World Bank, Washington, DC.

World Bank, 2012. Turn Down The Heat: Why a 4°C Warmer World Must Be Avoided. 106 pp. November 2012.

Chapter | Seven

# Conclusions

## 7.1 BACKDROP

There is enough food produced in the world to ensure that each of us is well-nourished. Yet there is a common mismatch between its availability on the one hand and the ability to access it on the other, this affecting many people, especially in the developing world. As a result, in 2015 an estimated 795 million of us from the total world population of over 7 billion were hungry (undernourished in terms of dietary energy). Of these 7 billion, many are also (or instead) suffering from deficiencies of one or more minerals or vitamins, a type of undernourishment which UNICEF has called the "silent urgency"—for example, more than 2 billion people are anemic, many due to iron deficiency. By contrast, 1.5 billion of us globally are malnourished due to consuming an *oversufficiency* of dietary energy, almost 1 billion of these in developing countries. Both deficiency and oversufficiency can lead to one or more metabolic dysfunctions, hence morbidity and even death. To be food-secure does not necessarily mean that an individual is nutritionally secure.

Most of the undernourished live in developing countries, many of which have a shortfall of grain and other foodstuffs grown within their borders to feed their populations. Yet, although food and nutrition insecurity is a reality or threat to those who are poor and marginalized, it is a threat to the food security and social institutions of us all.

Food insecurity, in which people are either currently undernourished or vulnerable to becoming so, may be short-term (transitional) or long-term. An example of short-term is during the "hunger season", before the next harvest is due, with that from the previous season having been exhausted or nearly so. By contrast, long-term food insecurity can occur, for instance, when the annual rains fail twice or more in a row.

To be food-secure, food must be available, access to it must be assured (physical, economic and social), it needs to be prepared well and consumed adequately, be nutritious and free from contamination. All these components need to be in place simultaneously, and in a stable not intermittent way. Some communities and individuals within them are more vulnerable to food insecurity and/or nutritional insecurity than others, and this is often measured by survey. However, food security is but one element of *livelihood security*,



Food Security in the Developing World. DOI: http://dx.doi.org/10.1016/B978-0-12-801594-0.00007-5

and indicators of the former should not be interpreted independently of a good understanding of the latter (this is discussed further in the introduction to the Case Studies on the book's companion website).

## 7.2 CAUSES OF FOOD INSECURITY

Causes of food insecurity include poverty, engrained social norms and insufficient awareness of the complex issues involved; environmental degradation due to population pressure, and climate change; land tenure insecurity; water insecurity; food price hikes and price instability; conflict; weak enabling environment, in public and private sectors; predisposition of the community to disease and intestinal afflictions; large-scale land lease by an investor, or by a sovereign government of one country in another country; and, large areas of arable land set aside for biofuel production. An individual, family or community may be beset by one or more of these causes.

Insufficient economic access to nutritious food is one of the main reasons why 795 million people were still undernourished in 2015, at the end of the MDG monitoring period. Poverty of *income* means that many people cannot eat sufficient food to stave off hunger. Increased volatility in agricultural markets has underlined the acute vulnerability of the poorest to even a small additional shock, for example, a sudden price rise in food or fuel, and a poor harvest. Poverty of *knowledge and awareness* in a society can lead to undernutrition in the form of insufficient protein, minerals or vitamins for a range of reasons—because these items are not sought or bought, that their intake is subject to restrictive feeding practice social norms pertaining to infants or post-pregnancy women (as in Lao PDR, Belize or northern Nigeria, for example), that their absorption in the digestive tract is sub-optimal because of intestinal afflictions resulting from poor hygiene and contaminated water, and/or that their use in the body is insufficient because of one or more diseases which prevent the nutrients being best used at cellular level. Affliction by both intestinal parasites and disease inhibits appetite, which also fosters undernutrition.

## 7.3 WAYS TO RELIEVE CURRENT FOOD AND/OR NUTRITION INSECURITY

Ways to relieve current food insecurity include implementing one or more coping strategy by the individual or community concerned. If/when these fail, humanitarian aid is an option, from one or more development partner—government, NGO or international organization, though many in real need do not attract timely attention and their needs go untended. There are so many simultaneous emergencies worldwide, it is not surprising that the combined efforts of international and charity organizations do not reach all those in need. Though having the potential to save lives and relieve suffering, humanitarian aid is unsustainable. Better by far if such assistance can be merged with development assistance, structured to build resilience in the community against future food and nutrition insecurity, such as education in enhanced primary healthcare and diet (see Section 4.7.4) and better care of the environment (see Case Study 1 on the book's companion website).

## 7.4 COLLATERAL EFFECTS OF FOOD INSECURITY

In addition to hunger and undernutrition outcomes of food insecurity, the latter can lead to political and personal insecurity, nationally and regionally. Food riots by the urban poor in much of the developing world downstream of the food price hike of 2008 showed how political security and national equilibrium can be thrown into disarray (see Case Study 9). Nor do hungry people keep their national borders, many becoming refugees in the quest to find food and a stable environment in which to farm, or earn so they may buy food. Often the influx of these refugees into a host country results in problems there, such as Syrian refugees in Lebanon, Ecuador struggling under the influx of Columbians (see Case Study 4, on the book's companion website) and economic refugees from elsewhere in Central America being employed in plantations in Belize rather than Belizean nationals (see Case Study 3).

In Case Study 5 on Amazonia and the La Plata Basin, attention is drawn to a nexus of interlinked securities which are under threat, not just food and nutrition security—water, energy, land tenure and use, livelihood and health security as well as regional stability, all compromised by human behavior, driven by lust for money and power, or politicians caught between the competing imperatives to bring development and wealth to their country or protect the integrity of the environment's ecological services. This conflict of interests is playing out across the planet, not just in Amazonia. All these threats are a direct or an indirect outcome of increasing population pressure, and resulting competition for the planet's finite natural resources.

## 7.5 WHAT OF THE FUTURE?

As stated at the end of Chapter 1 "Introduction", the challenge of feeding the world over the 35 years from 2015 to 2050 is enormous. The global food system as currently practised works for the majority, but not for those who are extremely poor and marginalized, most of whom live in developing countries. This book has presented an analysis of where we are now, and what is being done to mitigate current food insecurity and prevent future food insecurity.

The good news is that the numbers and proportion of food and nutritionally insecure people on the planet are reducing. This is because many of the various causes are in the process of being removed. A notable exception is the escalation of conflict, the world community overwhelmed by the numbers of IDPs and refugees, with individual rations UNHCR is able to provide being reduced so that more of the displaced can at least get something.

Apart from the tragedy of the displaced, there is now a realization that the current global food system is unsustainable, that doubling production to feed twice the number of people by 2050 may not be possible in a business-as-usual way, that inaction is not an option, and that the rate of change in policy, implementation and attitude has to change up a gear or two, in the interest of creating a more equitable world.

Looking beyond 2050, one has to hope for a stabilization of the global population number. Rights issues, including that of food security and adequate nutrition, and removal of poverty, are high on the development agenda these days. Fortunately, all these changes are consistent with common sense, and we have to trust in that, the trust espoused by the poet T.S. Eliot and his mystic mentor, in the poem *Little Gidding*:

> *. . .. And all shall be well and*
> *All manner of thing shall be well*
> *By the purification of the motive . . ..*

Yet "purifying the motive" (in this case, removing global food and nutrition insecurity permanently) will in itself be insufficient. Some "teeth" are needed. In developing countries, seeking to remove one or more causes of food insecurity, decrease vulnerability and build resilience against decline or relapse, involves actions as discussed in Chapter 4 "Mitigation of Current Food Insecurity" and Chapter 5 "Prevention of Future Food Insecurity", and Section 7.3. Anticipating crises by assessing risks and vulnerability is fundamental to this—early warning systems are vital, as also formulating national food security strategies to encourage donor partner buy-in. Various approaches are examined in this book to improve the food availability and food access components of food insecurity, individually and in concert.

The *availability* component of food security includes the work of plant breeders/agronomists/physiologists, improving the management of livestock and support to artisanal fisheries, and the imperative of reducing global food loss and waste. The improved *access* component includes enhancing the wealth creation enabling environment, through encouraging employment and investment, improving public sector policies for business and social protection. Addressing both *availability* and *access* in tandem includes the option of food reserves, and better identification of investment and trade opportunities. The especial challenge of achieving the elusive *stability* component of food security has proven the hardest task of all.

Nutrition security can be tackled sustainably in part by biofortification of crops, through the work of breeders and biochemists to create a *supply* of more nutritious cultivars, and anthropologists to create the *demand* and gently foster modern attitudes in society toward gender equity and social norms related to dietary intake, domestic food preparation and feeding practices, which are in the interests of not only infants and women, but the men in the household. Caring practices, especially of children, are often sub-optimal, such as in the highlands of Lao PDR (see Case Study 6), and northern

Nigeria. Yet it is in northern Nigeria too where an exemplary government-owned social transfer mechanism is in place to address the results of this insufficiency of awareness and care at household level; the two coexist side by side (Section 4.7.4). Nutrition awareness campaigns are much needed, while dietary supplementation and fortification initiatives by governments and their development partners must continue to address current protein-calorie and micronutrient undernutrition.

Assuring adequate food security and nutrition requires a coherent global joined-up multi-sectoral multiagency *commitment* to be in place, and an enabling environment to ensure that a number of variables are optimized simultaneously and consistently at national and sub-national level (see Section 3.6.1). Acting now in a pre-emptive way is better than "firefighting" later, when beneficial options have been reduced. There is an overwhelming case for all stakeholder groups—governments, farmers and fishers associations and other CBOs, the commercial private sector, international organizations and NGOs to work together like never before, to build on what is rational, sustainable and possible. A message emerging from the 2014 report on *The State of Food Insecurity in the World* is that with the requisite political *commitment*, accelerated, substantial and sustainable hunger reduction is possible. This has to be well-informed by a sound understanding of national challenges, relevant policy options which specifically target the poor and women, broad participation and lessons learned from other experiences. The word "commitment" has been mentioned throughout this book. It is that which has resulted in the MDG 1c target being achieved, and it is that across the board which will ensure that the numbers of food- and nutritionally-insecure numbers can be reduced to zero.

Global food security and nutrition security measures must be prioritized to foster sustainable use of our land and water resources in order to harvest our food to the level needed, a level which will double over the next 35 years, with scant sustainable means of increasing the area of arable land and availability of fresh water. Increase in productivity per unit of land and water, and decrease in food losses and waste are the ways to go on the supply side, while stemming the burgeoning human population on the demand side, based on the success of countries like Rwanda.

Good and accountable governance in general must be in place, not least that pertaining to rights-based food security *per se* and the imperative of slowing the speed and intensity of climate change; universal provision of clean water and proper sanitation; good health must be assured in general, meaning good primary healthcare, counseling and medical treatment centers in numbers sufficient for the population density; adequate educational facilities are needed for both girls and boys to enable enlightened social "norms" to be tabled and discussed, and for well-informed decisions to be taken by caregivers and heads of household on matters related to health and nutrition; poverty must be tackled, and one way in which this can come about is through wise investment in market-led enterprises which generate decent jobs and wealth. Cross-cutting issues include the necessity of gendering food security and providing secure land tenure for farmers, and encouraging family planning to improve the chance of each child having sufficient resources to survive childhood and prosper beyond it.

Alongside these priorities there needs to be reform of some trade and subsidy policies and attendance to all the related imperatives, such as preventing contamination of land and water resources with our industrial and human waste. These changes cannot be brought about relying solely on the wisdom of bureaucrats and politicians; stronger drivers may be a consensus of public opinion, for which the worldwide web, social and international media will likely play significant parts—citizen empowerment—both to formulate sound evidence-based plans, and to monitor and evaluate their implementation.

Of all the advisories written on the way forward, there is one which the current author would recommend to his readers to supplement this book and its companion website, namely *The Future of Food and Farming* published by the British government, which charts a dozen priorities at policy and implementation levels (The Government Office for Science, 2011). These policy priorities stress the importance and urgency of investment in people and infrastructure, agricultural and fishery extension and research, and reclamation of degraded land once used for agriculture, with proper incentives in place for those who farm and fish. The sustainability aspect of food production on land and in water is also stressed, and the participation of all stakeholders, involving Public-Private Partnerships and international organizations.

## 7.6 TOP PRIORITY—WOMEN'S EMPOWERMENT AND GENDER TRANSFORMATION

If the current author were to select just one cause of food insecurity and undernutrition that most warrants address and "kicking permanently into touch", the one which in his view is at the apex of the pyramid of all the others, attention to which would automatically dispel many other causes ... it is the need for gender to be mainstreamed, for women to be better empowered and become active agents of change to dispel poverty (see Section 6.2).

With women empowered, they would be able to limit the number of children they have so each has a better chance of success in life, and there would be fewer mouths that they would have to feed by their labors in the field; girls would not marry early without first receiving a good education and training in life skills, and so would be better able to earn an income even while looking after their (fewer) children—and this income would be invested in a better diet for the under 5 years of age, and in the land to increase its productivity; women would participate in the fundamental decisions made in the home, not just the peripheral ones; with better-educated women at home, there would be better standards of health and hygiene, less diarrhea in the under five years of age and fewer child deaths ... and happier husbands.

One of the most impressive pieces of research which the current author located during the preparation of this book was by Silvia Kaufmann on the determinants of undernutrition in Lao PDR. One of her main conclusions was that decentralized capacities and responsibility for decision-making, according to the law of subsidiarity, is the most effective modality to improve nutrition status in Lao rural communities. This went hand in hand with the finding that *better education* and *social status of women*, through improved "nutrition-supportive" decisions and behavior, mitigated the negative effects of food insecurity, poverty or other factors associated with a household's location, such as remoteness and limited access to social services or markets; these two factors determined nutritional status more than did any other variable in a multifactorial analysis (see Case Study 6, para 6.4 on the book's companion website).

With better-educated women in public life, there would likely be fewer wars. Experience and studies show that what "peace" means, and how to achieve and keep it, are usually seen in different perspectives by men and women. So, educating and empowering women may lead to that most intransigent cause of food insecurity—conflict—being mitigated, through better local and national governance, the achievement and stability of food security thereby enhanced.

It follows that development and humanitarian aid workers in the field of food and nutrition security need skills not only in technical disciplines but social and anthropological skills as well. Change in a society can be slow, but creating facts on the ground has proven to be a winning strategy, for women in Afghanistan for instance. Social norms can change downstream of actual change on the ground. Any change that happens is what is anthropologically possible, led by those "local people" who command respect and serve as agents of beneficial change, now sometimes called "champions".

In 2011 the current author worked with Her Excellency Husn Banu Ghazanfar, the Minister of Women's Affairs in Afghanistan, in formulating the country's first rural development report. He would like to cite her as one such "champion", who together with Afghan women who bravely and successfully campaigned to become members of the parliament in Kabul, have achieved significant changes on the ground against all the odds in a traditional patriarchal society. One woman, one man, really can make a difference, for the better.

The example of the man who started *haat* bazaars in the eastern Himalayas of Nepal was mentioned in Section 5.4.1. Starting off as a barter system it then became monetized, with volume of trade and living standards in villages boosted, women taking the lead role in marketing. Also in the lower mountain slopes of Nepal, the current author made friends with the man in Arghakhanchi district of western Nepal who introduced ox draft cultivation to the hillsides. Initially scorned by the community, Mr Jiwali persevered, leading by example, and gradually the idea took hold, with larger areas cultivated and more quickly. No longer laughed at, he is now hugely respected as an innovative visionary leader. Ox draft tillage and weeding on *bari* rain-fed terraces of the steep slopes is the new "norm", conducted by men and thereby saving much of the onerous burden of digging by hoe, much of that done by women. Creating other new win-win "norms" and opportunities in the interests of better food and nutrition security in the developing world is the way to go ... with a one-way ticket. The achievement of MDG 1c targets in so many developing countries has surprised many doubters in the developed world. None of us can afford a return ticket to 1990.

## REFERENCE

The Government Office for Science, 2011. Foresight. The future of food and farming. Final Project Report. London. 211 pp. <https://www.gov.uk/government/publications/future-of-food-and-farming>.

# Epilogue: A Rude Awakening

In anyone's life there are watershed moments in which one finds oneself on a near upwardly vertical learning trajectory. For the author, one such moment was at the bedside of a child in the government malnutrition clinic at Mulago hospital, Uganda. The child died in his presence, the first child he had ever seen die. The mother, without touching her son burst into loud wailing and fled the bedside, leaving him and the Ugandan staff nurse with her dead child. The nurse abused the mother as she left because she had responded so. Strangely, as a non-Ugandan, the author gently chastised the nurse that she should be more considerate of local custom. Looking back on it, how superfluous the three separate behavior patterns were ... one more child had died, for an inglorious and preventable reason for which we should all be ashamed and take responsibility, whoever we are, wherever we are.

As a rookie lecturer in human nutrition to agricultural students at Makerere University during General Idi Amin's regime, this was a lesson for the author on how poverty is a more silent stalker of life than the gun, and how closely poverty, conflict, hunger and undernutrition are conjoined. He still meets his former women students on the streets of Kampala sometimes, who tell him that the Mulago practicals were the best thing that happened to them in their undergraduate life. An afternoon in which they learned that the sciences of agriculture and medicine are inextricably linked to each other, and to the continuum of health, morbidity and death. This is analogous to the great advances made in the sciences when two erstwhile separate fields have been bought together in the same research program, allowing commonalities and synergies to be better understood and exploited, like biochemistry and biophysics in the treatment of cancer, or thermodynamics and mechanics in advancing propulsion engineering.

With scientists telling us that the world has embarked on the sixth phase of mass extinction of species (see Section 3.3) we must shake off any feeling that we can continue in a business-as-usual modality, to expect our food, water and energy supplies to be assured *ad infinitum*. Never has a more urgent call to action been warranted than now. We all need to work together seamlessly and urgently, with wiser political leadership than currently we often have, marshalling all the technical, economic and anthropological expertise in a multisectoral and coherent way, to ensure sustainable food security for us and our descendants in perpetuity.

The reader may now wish to consult the companion website associated with this book, comprising 10 case studies providing snapshots to elaborate on key concepts and variables that have been discussed in this book. Readers requiring yet more detail can refer to the reference lists provided, for both the book and the companion website.

# Index

Note: Page numbers followed by "*b*," "*f*," and "*t*" refer to boxes, figures, and tables, respectively.

## C

## D



## E



## F

## G

## H



## I



## J



## K



## L

## U



## V



## W



## Z

Printed in the United States
by Baker & Taylor Publisher Services